Ecology

Ecology

Avery Nash

MURPHY & MOORE
www.murphy-moorepublishing.com

Published by Murphy & Moore Publishing,
1 Rockefeller Plaza,
New York City, NY 10020, USA
www.murphy-moorepublishing.com

Ecology
Avery Nash

International Standard Book Number: 978-1-63987-176-6 (Hardback)

Cataloging-in-Publication Data

Ecology / Avery Nash.
p. cm.
Includes bibliographical references and index.
ISBN 978-1-63987-176-6
1. Ecology. 2. Biology. 3. Environmental sciences. I. Nash, Avery.
QH541 .E26 2022
577--dc23

TABLE OF CONTENTS

PREFACE

The branch of biology which deals with the spatial and temporal patterns of the distribution of organisms on Earth is termed as ecology. It also focuses on the interaction between various organisms and their environment. The study of the impact of various biotic and abiotic factors on the ecosystems is also conducted within this field. The biotic components consist of living factors such as bacteria, animals, plants, birds, fungi, etc. The non-living physical and chemical factors affecting ecosystems are termed as abiotic components. Sunlight, water, air and soil are some of the major abiotic components. The discipline can be broadly categorized into global ecology, landscape ecology, population ecology, community ecology, ecosystem ecology and organismal ecology. It finds extensive application in the fields of wetland management, conservational biology, community health and economics. The book studies, analyses and upholds the pillars of ecology and its utmost significance in modern times. It brings forth some of the most innovative concepts and elucidates the unexplored aspects of ecology. This book will serve as a valuable source of reference for those interested in this field.

A short introduction to every chapter is written below to provide an overview of the content of the book:

Chapter 1 - Ecology is a branch of biology that deals with organisms, their physical environment and their interactions. It is further divided on the basis of organisms, biomes and scales. It is studied at different levels like population ecology, community ecology, ecosystem ecology, system ecology, landscape ecology, etc. This chapter serves as an introduction to ecology and its related aspects in a clear and concise manner; **Chapter 2 -** An ecosystem can be defined as a community or a group of living organisms within a geographical area that live and interact with each other and the environment. An ecosystem consists of both living and non-living components. Ecosystems, their different types and components are discussed in this carefully written chapter to provide a better understanding of the subject; **Chapter 3 -** The living components of an ecosystem can be organized into producers and consumers. They form an intricate association with each other through food chains and food webs. Energy flows through the ecosystem through these chains in a process called the energy cycle. The topics elaborated in this chapter will help in gaining a better perspective of energy flow in ecosystems; **Chapter 4 -** The habitat and the physical environment along with the various living components of an ecosystem, biodiversity and its related cycles form the essential aspects of ecology. This chapter goes through each and every aspect of ecology in a detailed manner to provide the reader with an extensive understanding of the topic; **Chapter 5 -** Biogeochemical cycle refers to any of the natural pathways by which essential elements of living matter are circulated through the various biotic and abiotic parts of an ecosystem. Water cycle, carbon cycle, nitrogen cycle, oxygen cycle, phosphorus cycle, sulfur cycle, etc. are all part of the biogeochemical cycle. This chapter deals with the various biogeochemical cycles in a comprehensive manner for the benefit of the reader.

Finally, I would like to thank my fellow scholars who gave constructive feedback and my family members who supported me at every step.

Avery Nash

Chapter 1

Basics of Ecology

Ecology is a branch of biology that deals with organisms, their physical environment and their interactions. It is further divided on the basis of organisms, biomes and scales. It is studied at different levels like population ecology, community ecology, ecosystem ecology, system ecology, landscape ecology, etc. This chapter serves as an introduction to ecology and its related aspects in a clear and concise manner.

Ecology is a branch of science, including human science, population, community, ecosystem and biosphere. Ecology is the study of organisms, the environment and how the organisms interact with each other and their environment. It is studied at various levels, such as organism, population, community, biosphere and ecosystem. An ecologist's primary goal is to improve their understanding of life processes, adaptations and habitats, interactions and biodiversity of organisms.

Biotic and Abiotic Factors

The main aim of ecology is to understand the distribution of biotic and abiotic factors of living things in the environment. The biotic and abiotic factors include the living and non-living factors and their interaction with the environment.

Biotic Components

Biotic components are living factors of an ecosystem. A few examples of biotic components include bacteria, animals, birds, fungi, plants, etc.

Abiotic Components

Abiotic components are non-living chemical and physical factors of an ecosystem. These components could be acquired from the atmosphere, lithosphere and hydrosphere. A few examples of abiotic components include sunlight, soil, air, moisture minerals and more.

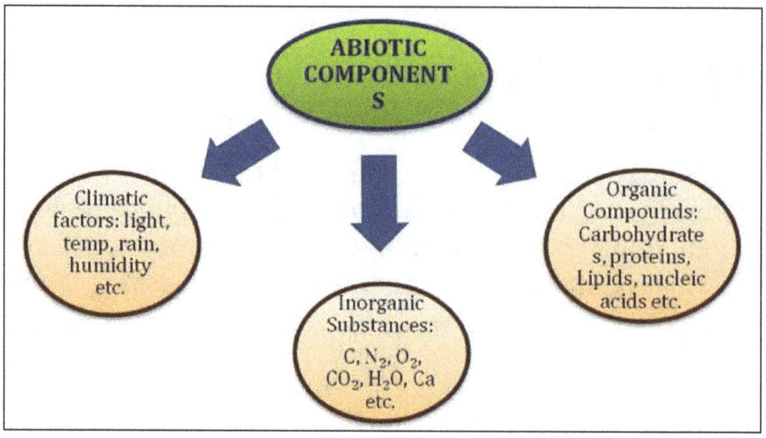

Living organisms are grouped into biotic components, whereas non-living components like sunlight, water, topography are listed under abiotic components. The following reasons explain the importance of ecology:

- Conservation of Environment: Ecology helps us to understand how our actions affect the environment. It shows the individuals the extent of damage we cause to the environment. Lack of understanding of ecology has led to the degradation of land and the environment. It has also led to the extinction and endangerment of certain species. For e.g., dinosaurs, white shark, mammoths, etc. Thus, the study of the environment and organisms helps us to protect them from any damage and danger.
- Resource Allocation: With the knowledge of ecology, we are able to know which resources are necessary for the survival of different organisms. Lack of ecological knowledge has led to scarcity and deprivation of these resources, leading to competition.
- Energy Conservation: All organisms require energy for their growth and development. Lack of ecological understanding leads to the over-exploitation of energy resources such as light, nutrition and radiation, leading to its depletion. Proper knowledge of ecological requirements prevents the unnecessary wastage of energy resources, thereby, conserving energy for future purposes.
- Eco-Friendliness: Ecology encourages harmonious living within the species and the adoption of a lifestyle that protects the ecology of life.

Examples of Ecology

Following are a few examples of ecology:

- Human Ecology: It focuses on the relationship between humans and the environment. It emphasizes the impact human beings have on the environment and gives knowledge on how we can improve ourselves for the betterment of humans and the environment.
- Niche Construction: It deals with the study of how organisms alter the environment for the

benefit of themselves and other living beings. For Example:, termites create a 6 feet tall mound and at the same time feed and protect their entire population.

Ecology by Level

Community Ecology

Community ecology is the study of the organization and functioning of communities of organisms. As populations of species interact with one another, they form biological communities. A community of organisms consists of all the interacting populations of the species living within a particular area or within a particular habitat. Community ecology also studies the relationships of the members of a community to their environment. Community ecology is usually subdivided according to habitat or biome. Typical habitats include forest, grassland, desert, and stream or lake environments.

The Trophic Pyramid

All biological communities have a similar structure called a trophic pyramid. Each pyramid contains four or five levels. Food energy is passed from one level to the next along a food chain. Since energy is lost to heat at each level in the pyramid, it takes many organisms at a given trophic level to support those in the next level up. The base of the pyramid in every biological community is composed of species called autotrophs, organisms that harvest sunlight (or in rare cases, heat) directly through photosynthesis (or chemosynthesis). All other organisms in the pyramid are called heterotrophs.

A food chain typically contains four or five links, from autotrophs, through grazers and other herbivores, then culminating with a carnivore as top predator. Many animals, however, eat more than one species. Also, animals may eat different foods at different stages of their growth. Many animals eat both plants and other animals and therefore feed at more than one trophic level. Consequently, food chains are usually interconnected into highly complex food webs.

In addition to eating one another, species also compete for resources and interact in other ways within a community. Nontrophic relationships between species are as important as food chains and food webs in shaping the organization of biological communities.

Ecological Succession

Through the process of ecological succession, communities are constantly changing. Disturbances to communities may be local, such as a tree falling and opening the canopy to allow more sunlight, or widespread, such as fires and storms. Whether local or general, each disturbance creates an opportunity for a new species to colonize that region. These new species can alter the biological structure of the community and create an environment that is suitable to other new species. By this process, the community evolves over time.

In some environments, succession eventually produces a stable community dominated by a small number of species. This is called a climax community. The web of biological interactions has become so intricate and interconnected that no other species can successfully compete for food resources. In other environments, small disturbances produce communities that are a diverse mix of species. Some tropical forests contain hundreds of thousands of species within a square kilometer. When a tree dies and falls, the dense canopy is opened and new space is available for different species to take root. Some coral reefs contain thousands of different species, and whichever species is able to rapidly colonize a new disturbance patch will be successful.

Ecological Niches

The way of life of an organism is shaped by its environment and by its interactions with other organisms through the processes of evolution. The role an organism plays in its relation to other species and its environment is known as its ecological niche. The niche of an organism includes what it eats, how it obtains food, where in the environment it lives, what temperature it prefers, how much light it can tolerate, and many other factors.

Guilds

Some similar species have evolved strategies that allow them to allocate resources in a way that avoids competition. For example, different species of warblers that prey on the same species of insects may forage at different levels in the same trees. A group of organisms that share a common food resource is called a guild. Guild members may have strong interactions with each other but only weak interactions with other members of the community. In the American Southwest, birds, rodents, and ants constitute a guild that competes for the same seeds. Whereas birds exploit temporary patches of seeds, rodents and ants are permanent residents. Ants generally take smaller seeds than rodents. In East Africa, communities of animals form a guild of grazers. First, elephants and buffalo eat the tall, coarse grasses and then move on. They are able to consume large quantities of this low-nutrition food source. Zebras follow along behind the elephants, reducing the plant biomass even more. The zebras are followed by a still smaller animal, the wildebeest, which selects among the lower growing plants that remain after the zebras have fed. Finally, the smallest grazers, such as Thompson's gazelles, are able to reach the young, protein-rich sprouts of grass missed by the wildebeest.

Interactive Relationships

The interactive relationships that arise between populations of different species form the interactive web of communities. These interactions range from antagonistic to cooperative and have positive, negative, or neutral effects on the species involved. In antagonistic relationships the interaction is detrimental to individuals of either one or both species; in commensal relationships (commensalism) one species benefits while the other remains unaffected; and in mutualistic relationships (mutualism) both species benefit. The organization and stability of biological communities results from the mix of these different kinds of interaction.

There are many different kinds of interspecific interactions within an ecological community. These relationships between species are not static; they evolve as natural selection continually shapes and reshapes them. The complex relationships between prey and predators, for example, are snapshots of one instant during the evolution of interactions. As interactions between species evolve, the nature of relationships may shift. Nineteenth-century British naturalist Charles Darwin called this ever-changing mix of species and their interactions the "entangled bank" and stressed its importance in the evolutionary process. While antagonistic relationships, such as predator-prey or parasite-host, are the most dramatic kinds of relationships, other forms of interaction such as mutualism or commensalism are just as important.

Mutualism

This is a relationship where both participants in the interaction receive benefit. For example, plants are hosts for insects that pollinate them or eat their fruit and for microorganisms that attach themselves to their roots. Mutualistic associations between animals and microorganisms are an important part of the structure of communities. Most animals rely on the microorganisms in their gut to properly digest and metabolize food.

Parasitism

This is possibly the most common way of life in nature. Parasitic organisms may account for half of all living species. The majority of species of wasps are tiny parasites that lay their eggs on a specific host organism. Some wasps are parasitic on plants, some are parasitic on insects, and some even parasitize other wasps! The larvae hatch and burrow into the host species. As it is consumed from the inside out, the host species survives long enough to allow the larvae to mature.

Antagonism

Antagonism is a form of relationship where one species benefits and the other are harmed. Grazing, parasitism, and predation are examples of antagonistic relationships. While we generally think of grazers as large herbivores, a grazer is defined as any species that moves from one organism to another, feeding on part of each without actually killing it outright. Grasshoppers are grazers that jump from plant to plant, chewing a portion of the leaves of each one they visit. Some caterpillars are grazers that crawl from one plant to another during development rather than remain as parasites on an individual plant. The grazing lifestyle differs from the parasitic lifestyle in a few important ways. Individuals can vary their diets with different foods. Also, because grazers do not remain attached to a single individual for long periods, their victims do not have time to develop induced specialized defenses, such as an immune response that a host can develop against a parasite.

Predation

This form of relationship differs from both parasitism and grazing. In predation, the victims are killed and often consumed immediately. Predators therefore differ from parasites and grazers in their effects on the dynamics of populations and the organization of communities. As with parasitism and grazing, predation is an interaction that has arisen many times in many taxonomic groups worldwide. Competition is an important form of interaction in communities in which neither species benefits. In competitive interactions, species evolve either to avoid each other, to tolerate the presence of the other, or to aggressively exclude the other.

Species compete for almost every conceivable kind of resource. Birds compete for nesting sites.

Male birds compete for preferred sites to defend as territories for attracting females. While species compete for many resources at the same time, there is often a single resource, called the limiting resource that is in scarce supply. This resource restricts the growth of each species. In deserts, water is often the limiting resource.

Commensalism

In this kind of interaction, one species benefits and the other is unaffected. For example, cattle egrets forage around the feet of cattle. The grazing behavior of the cattle stirs up many small insects and other arthropods that the cattle egrets eat. The cattle egrets receive a benefit, but there is no indication that the cattle are affected in any way.

The richness and ubiquity of interactions among populations of organisms demonstrate that the characteristics of all species have been influenced by the interaction with other species. Species have coevolved with each other. Predators have evolved along with their prey. Parasites evolve with their hosts. Rarely is only one interaction responsible for the evolution of a species, however. More common is a sort of diffuse coevolution where the traits of a single species are influenced by interactions with many other species. Such diffuse coevolution may prevent a sort of evolutionary "arms race," where predator and prey become ever faster or stronger.

Ecosystem Ecology

An ecosystem is a community of living organisms and their interactions with their abiotic (non-living) environment. Ecosystems can be small, such as the tide pools found near the rocky shores of many oceans, or large, such as the Amazon Rainforest in Brazil.

(a) (b)

(a) tidal pool ecosystem in Matinicus Island in Maine is a small ecosystem, while the (b) Amazon Rainforest in Brazil is a large ecosystem.

There are three broad categories of ecosystems based on their general environment: freshwater, ocean water, and terrestrial. Within these broad categories are individual ecosystem types based on the organisms present and the type of environmental habitat.

Ocean ecosystems are the most common, comprising 75 percent of the Earth's surface and consisting of three basic types: Shallow Ocean, deep ocean water, and deep ocean surfaces (the low depth areas of the deep oceans). The shallow ocean ecosystems include extremely bio diverse coral reef ecosystems, and the deep ocean surface is known for its large numbers of plankton and krill (small crustaceans) that support it. These two environments are especially important to aerobic respirators worldwide as the phytoplankton perform 40 percent of all photosynthesis on Earth. Although not as diverse as the other two, deep ocean ecosystems contain a wide variety of marine organisms. Such ecosystems exist even at the bottom of the ocean where light is unable to penetrate through the water.

Freshwater ecosystems are the rarest, occurring on only 1.8 percent of the Earth's surface. Lakes, rivers, streams, and springs comprise these systems; they are quite diverse, and they support a variety of fish, amphibians, reptiles, insects, phytoplankton, fungi, and bacteria.

Terrestrial ecosystems, also known for their diversity, are grouped into large categories called biomes, such as tropical rain forests, savannas, deserts, coniferous forests, deciduous forests, and tundra. Grouping these ecosystems into just a few biome categories obscures the great diversity of the individual ecosystems within them. For example, there is great variation in desert vegetation: the saguaro cacti and other plant life in the Sonoran Desert, in the United States, are relatively abundant compared to the desolate rocky desert of Boa Vista, an island off the coast of Western Africa.

(a) (b)

Desert ecosystems, like all ecosystems, can vary greatly. The desert in (a) Saguaro National Park, has abundant plant life, while the rocky desert of (b) Boa Vista island, Cape Verde, is devoid of plant life.

Ecosystems are complex with many interacting parts. They are routinely exposed to various disturbances, or changes in the environment that effect their compositions: yearly variations in rainfall and temperature and the slower processes of plant growth, which may take several years. Many of these disturbances are a result of natural processes. For example, when lightning causes a forest fire and destroys part of a forest ecosystem, the ground is eventually populated by grasses, then by bushes and shrubs, and later by mature trees, restoring the forest to its former state. The impact of environmental disturbances caused by human activities is as important as the changes wrought by natural processes. Human agricultural practices, air pollution, acid rain, global deforestation, overfishing, eutrophication, oil spills, and illegal dumping on land and into the ocean are all issues of concern to conservationists.

Equilibrium is the steady state of an ecosystem where all organisms are in balance with their environment and with each other. In ecology, two parameters are used to measure changes in ecosystems: resistance and resilience. The ability of an ecosystem to remain at equilibrium in spite of disturbances is called resistance. The speed at which an ecosystem recovers equilibrium after being disturbed, called its resilience. Ecosystem resistance and resilience are especially important when considering human impact. The nature of an ecosystem may change to such a degree that it can lose its resilience entirely. This process can lead to the complete destruction or irreversible altering of the ecosystem.

Food Chains and Food Webs

The term "food chain" is sometimes used metaphorically to describe human social situations. In this sense, food chains are thought of as a competition for survival, such as "who eats whom?" Someone eats and someone is eaten. Therefore, it is not surprising that in our competitive "dog-eat-dog" society, individuals who are considered successful are seen as being at the top of the food chain, consuming all others for their benefit, whereas the less successful are seen as being at the bottom.

The scientific understanding of a food chain is more precise than in its everyday usage. In ecology, a food chain is a linear sequence of organisms through which nutrients and energy pass: primary producers, primary consumers, and higher-level consumers are used to describe ecosystem structure and dynamics. There is a single path through the chain. Each organism in a food chain occupies what is called a trophic level. Depending on their role as producers or consumers, species or groups of species can be assigned to various trophic levels.

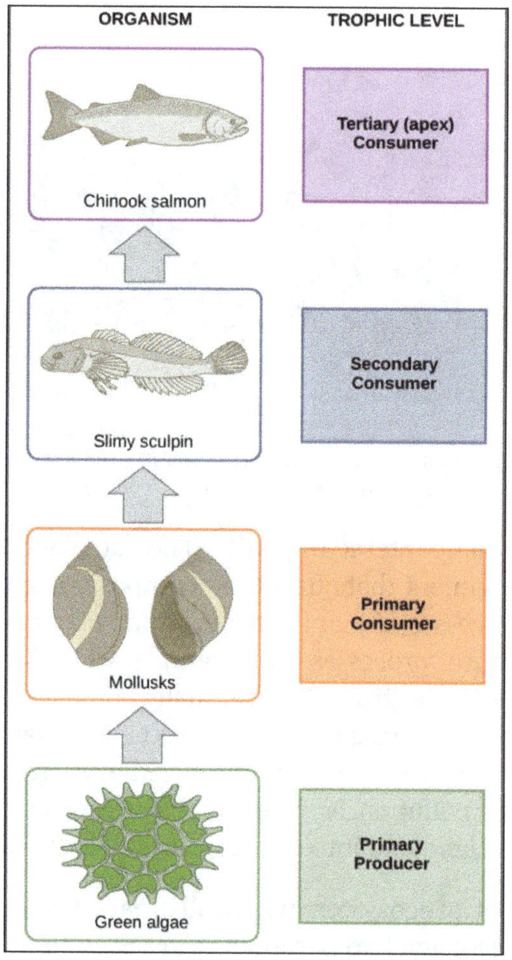

These are the trophic levels of a food chain in Lake Ontario. Energy and nutrients flow from photosynthetic green algae at the bottom to the top of the food chain: the Chinook salmon.

In many ecosystems, the bottom of the food chain consists of photosynthetic organisms (plants and/or phytoplankton), which are called primary producers. The organisms that consume the primary producers are herbivores: the primary consumers. Secondary consumers are usually carnivores that eat the primary consumers. Tertiary consumers are carnivores that eat other carnivores. Higher-level consumers feed on the next lower tropic levels, and so on, up to the organisms at the top of the food chain: the apex consumers. In the Lake Ontario food chain the Chinook salmon is the apex consumer at the top of this food chain.

One major factor that limits the length of food chains is energy. Energy is lost as heat between each trophic level due to the second law of thermodynamics. Thus, after a limited number of trophic energy transfers, the amount of energy remaining in the food chain may not be great enough to support viable populations at yet a higher trophic level.

The loss of energy between trophic levels is illustrated by the pioneering studies of Howard T. Odum in the Silver Springs, Florida, ecosystem in the 1940s. The primary producers generated 20,819 kcal/m²/yr (kilocalories per square meter per year), the primary consumers generated 3368 kcal/m²/yr, the secondary consumers generated 383 kcal/m²/yr, and the tertiary consumers only generated 21 kcal/m²/yr. Thus, there is little energy remaining for another level of consumers in this ecosystem.

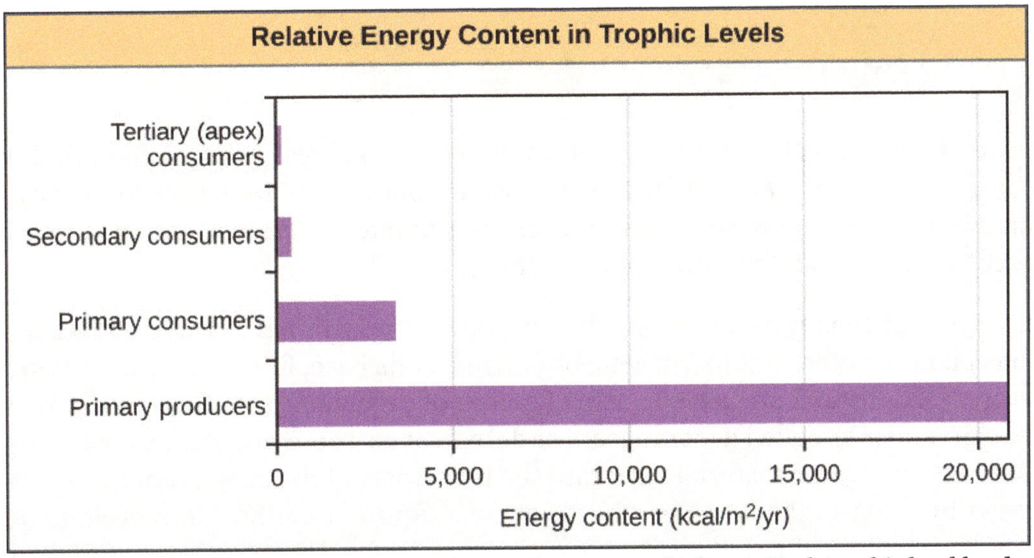

The relative energy in trophic levels in a Silver Springs and ecosystem is shown. Each trophic level has less energy available and supports fewer organisms at the next level.

There is a one problem when using food chains to accurately describe most ecosystems. Even when all organisms are grouped into appropriate trophic levels, some of these organisms can feed on species from more than one trophic level; likewise, some of these organisms can be eaten by species from multiple trophic levels. In other words, the linear model of ecosystems, the food chain, is not completely descriptive of ecosystem structure. A holistic model—which accounts for all the interactions between different species and their complex interconnected relationships with each other and with the environment—is a more accurate and descriptive model for ecosystems. A food web is a graphic representation of a holistic, non-linear web of primary producers, primary consumers, and higher-level consumers used to describe ecosystem structure and dynamics.

The above image of food web shows the interactions between organisms across trophic levels in the Lake Ontario ecosystem. Primary producers are outlined in green, primary consumers in orange, secondary consumers in blue, and tertiary (apex) consumers in purple. Arrows point from an organism that is consumed to the organism that consumes it. Notice how some lines point to more than one trophic level. For example, the opossum shrimp eats both primary producers and primary consumers.

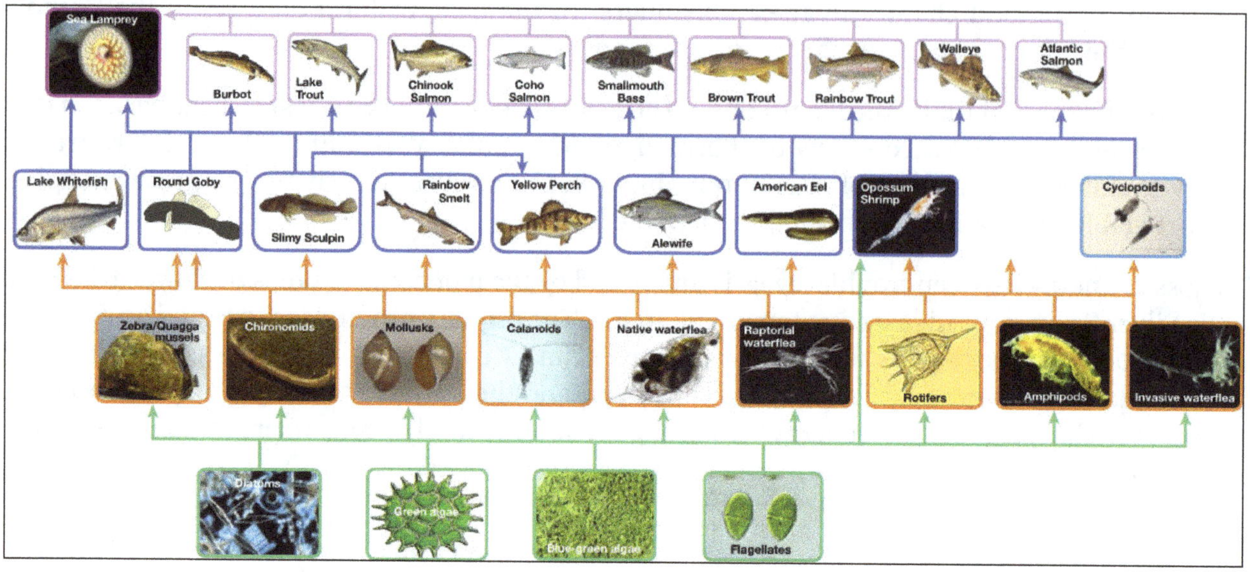

A comparison of the two types of structural ecosystem models shows strength in both. Food chains are more flexible for analytical modeling, are easier to follow, and are easier to experiment with, whereas food web models more accurately represent ecosystem structure and dynamics, and data can be directly used as input for simulation modeling.

Two general types of food webs are often shown interacting within a single ecosystem. A grazing food web has plants or other photosynthetic organisms at its base, followed by herbivores and various carnivores. A detrital food web consists of a base of organisms that feed on decaying organic matter (dead organisms), called decomposers or detritivores. These organisms are usually bacteria or fungi that recycle organic material back into the biotic part of the ecosystem as they themselves are consumed by other organisms. As all ecosystems require a method to recycle material from dead organisms, most grazing food webs have an associated detrital food web. For example, in a meadow ecosystem, plants may support a grazing food web of different organisms, primary and other levels of consumers, while at the same time supporting a detrital food web of bacteria, fungi, and detrivorous invertebrates feeding off dead plants and animals.

Systems Ecology

The analysis of how ecosystem function is determined by the components of an ecosystem and how those components cycle, retain, or exchange energy and nutrients. Systems ecology typically involves the application of computer models that track the flow of energy and materials and predict the responses of systems to perturbations that range from fires to climate change to species extinctions. Systems ecology is closely related to mathematical ecology, with the major difference stemming from systems ecology's focus on energy and nutrient flow and its borrowing of ideas from engineering. Systems ecology is one of the few theoretical tools that can simultaneously examine a system from the level of individuals all the way up to the level of ecosystem dynamics. It is an especially valuable approach for investigating systems so large and complicated that experiments are impossible, and even observations of the entire system are impractical. In these overwhelming settings, the only approach is to break down the research into measurements of components and then assemble a system model that pieces together all components. An important contribution of ecosystem science is the recognition that there are critical ecosystem services such as cleansing of water, recycling of waste materials, production of food and fiber, and mitigation of pestilence and plagues.

General systems theory is a science that studies the abstract properties of systems. It attempts to find the commonality in things as diverse as atomic particles, ecosystems, and political parties. The basic ideas were proposed in the 1940´s and 50´s by Ashby, von Bertalanffy, and others. They felt that the same techniques of interdisciplinary teamwork and mathematical analysis that had worked so well during World War II to produce weapons of destruction could help solve problems of health, starvation and social unrest as well. Their idea was that whenever objects interact, they must do so in certain well-defined ways. Study of new systems of objects should involve a determining which pattern of interaction the new system uses. It was hoped that patterns of interaction would be fairly limited so that they could be cataloged and studied as such. Except in certain narrowly defined fields (for example, queuing theory), this has not come to pass, although attempts have been made to produce such catalogs. General Systems Theory, while not yet succeeding in its ultimate goal, does offer a great deal to ecologists and natural resource managers:

- A structured world-view.
- A structured approach to problem solving.
- A theory of conceivable system behaviors.
- A set of powerful techniques for analyzing system behaviors.

Ecology is only a little older than General Systems Theory, starting at the end of the nineteenth century. Tansley introduced the term ´ecosystem´ but the concept is much older. Möbius wrote about an oyster reef as a ´biocoenosis´ and Forbes wrote about a lake as a ´microcosm.´ Ecology is the science of the interactions between organisms and their environment.

Despite over half a century of trying, it has not developed much beyond the stage of description. Many of the ´principles´ lay down by Odum and others in the early fifties are now being questioned. The strength of ecology still comes from its roots in Natural History, which goes back much further than 100 years. Ecology offers:

- A wealth of descriptive natural history data.
- Extremely complex interactions.
- A conceptual framework (ecosystem – community – population – individual).

The synthesis of these two sciences is called systems ecology. A systems ecologist is interested in interactions, and in whole system behavior. Mathematics often is involved, but the approach and attitude are more important than the tools. Science, including ecology, has tended to concentrate on small well-defined problems. Reductionism has been the rule. But General Systems Theory and systems ecology attempt to study systems as such, for example, organisms within their environment. Systems ecology involves:

- Constructing conceptual models of ecosystems.
- Statistical manipulation of data.
- Building dynamic models of ecosystems.
- Computer simulation.
- Applying systems analysis techniques to ecosystem models.
- Using all of the above to formulate new hypotheses and tests for hypotheses in the field.

Sects in Systems Ecology

A large number of approaches have already developed within ′systems ecology.′ Unfortunately, these various schools or sects are often on unfriendly terms, or at best seldom communicate with one another. The development of these sects was probably inevitable due to the diverse backgrounds of the scientists in the field, and due to the lack of firm laws in ecology. However, it should become apparent that all of these approaches have good points, and one should not scorn any of them. Ecosystems are too diverse, and ecologists are too small in number and too ignorant for any one group to have videos pornos gratis all the truth:

- Population-evolution approach: This sect uses Lotka-Volterra equations applied to communities. Important concepts include Kand r-selection, island biogeography, and competition or predation as organizing features of biological communities.
- Theoretical/Biomathematical Approach: This group uses simple equations to get mathematically tractable results and intuitive insights. Often optimization or optimal control is involved.
- Big Biology – Simulation Approach: Make big, complicated differential (or difference) equation models of big, complicated systems. The model forms the focal point of a large interdisciplinary team of researchers. The International Biological Program (IBP) spawned a number of these efforts, some of which are continuing today.
- Statistical-Manipulation-of-Data Approach: Take big data sets, such as those generated by the IBP, and apply factor analysis, principal components analysis, multivariate regression, etc. to gain insight into ecosystems, or to answer specific questions.
- Systems Dynamics Approach: Based on modeling work of Forrester and Meadows of MIT. This group produced the first world model, developed the DYNAMO computer simulation language, and a unique method of diagramming conceptual models.
- Linear Modeling and Systems Analysis: Attempt to use the mathematical power of linear systems theory and systems analysis techniques to model the major processes of ecosystems.
- Energy Flow Analysis: This is a group of ecologists who believe that energy is the common currency of all systems. By analyzing energy flow in both natural and man-made systems, they hope to discern much of the behavior of these systems.

- Management Modelers: This is a diverse group of people who use whatever technique they can to solve a problem presented to them in management of wildlife populations, parks, conservation preserves, etc. The driving principles here are to produce results that will improve management decisions.

Landscape Ecology

Landscape ecology is the study of the causes and ecological consequences of spatial pattern in landscapes. While there is no specific spatial extent that defines a landscape, most landscape ecologists are interested in large areas ranging from a few square kilometers to entire continents. Within landscapes it is usually possible to define a series of different ecosystem types occurring as patches within the greater landscape. For example, in an agricultural landscape the patches might be different fields, woodlots, hedgerows, buildings, and ponds. The goal of a landscape ecologist is to understand and describe landscape structure; how this structure influences the movement of organisms, material, or energy across the landscape; and how and why landscape structure changes over time.

A landscape's structure can be quantified by describing characteristics of patches, such as their number, size, shape, position, and composition. Landscape ecologists have defined measures to quantify each of these attributes. For example, a shape index has been defined as the ratio of the patch's perimeter to the perimeter of a circle the same area as the patch. A circular patch would have the value of 1, and as the patch became more convoluted in shape, its shape index would increase in value.

A landscape's structure has an important influence on various ecological processes occurring in the landscape. For example, consider two landscapes having equal areas of forest and agricultural land. In one landscape the forest is divided into many small patches, whereas in the other landscape the forested area occurs as one large patch. The more fragmented landscape will provide more habitats to those organisms that thrive at boundaries between two ecosystem types, whereas the less fragmented landscape will be better for those species that require larger areas of undisturbed forest. So, just knowing what percentage of the landscape is forest versus cropland is not sufficient to predict what species may occur; it is also important to know how the patches are distributed across the landscape.

Another example of how landscape structure can be important comes from studies of lakes within a forested landscape. The position of a lake within the landscape can be an important determinant of the lake's physical, chemical, and biological characteristics. Because water flows downhill, lakes that are lower in the landscape receive more water from streams and groundwater than lakes higher in the flow system, which receive most of their water from precipitation. Lakes higher in the landscape tend to be smaller, more dilute chemically, and have fewer species of fish than lakes lower in the landscape, even though all of the lakes in the landscape experience the same weather and are situated in the same geological substrate.

Landscape structure can change through natural geological or biological processes. Earthquakes, volcanoes, and landslides are examples of geological processes. The work of beavers building a dam to flood an area is an example of a biological activity that can change landscape structure. Human activity, such as the clearing of forest land for agriculture or the expansion of urban areas, has also caused significant changes in landscape structure. These changes in structure, whether caused by natural forces or by humans, can have significant impacts on the ecology of landscapes.

Ecology by Organisms

Animal Ecology

Animal ecology concerns the relationships of individuals to their environments, including physical factors and other organisms, and the consequences of these relationships for evolution, population growth and regulation, interactions between species, the composition of biological communities, and energy flow and nutrient cycling through the ecosystem. From the standpoint of population, the individual organism is the fundamental unit of ecology. Factors influencing the survival and reproductive success of individuals form the basis for under-standing population processes.

Two general principles guide the study of animal ecology. One is the balance of nature, which states that ecological systems are regulated in approximately steady states. When a population becomes large, ecological pressures on population size, including food shortage, predation and disease, tend to reduce the number of individuals. The second principle is that populations exist in dynamic relationship to their environments and that these relationships may cause ecological systems to vary dramatically over time and space. One of the challenges of animal ecology has been to reconcile these different viewpoints.

Populations depend on resources, including space, food, and opportunities to escape from predators. The amount of a resource potentially available to a population is generally thought of as being a property of the environment. As individuals consume resources they reduce the availability of these resources to others in the population. Thus, individuals are said to compete for resources. Larger populations result in a smaller share of resources per individual, which may lead to reduced survival and fecundity. Dense populations also attract predators and provide conditions for rapid transmission of contagious diseases, which generate pressure to reduce population size.

Changes in population size reflect both extrinsic variation in the environment that affects birth and death rates and intrinsic dynamics that result in oscillations or irregular fluctuations in population size. In some situations, the stable state may be a regular oscillation known as a limit cycle. Ecological systems also may switch between alternative stable states, as in the case of populations that are regulated at a high level by food limitation or at a low level by predators or other enemies. Switching between alternative stable states may be driven by changes in the environment.

Population Increase

In the absence of the effects of crowding, all populations have an immense capacity to increase. This capacity may be expressed as an exponential growth rate, which describes the growth of a population in terms of its relative, or percentage, rate of increase, like continuously compounded interest on a bank account. The constant r is often referred to as the Malthusian parameter. For a population growing at an exponential rate, the number of individuals (N) in a population at time t is $N(t) = (0)e^{rt}$ where N(0) is the number of individuals at time 0. Accordingly, the increase in a single time unit is e^r, which is the constant factor by which the population increases during each time period. The rate of increase in the number of individuals is then given by $dN/dt = rN$. The doubling time in years of a population growing exponentially is $t_2 = (\ln 2)/r$, or roughly 0.69/r.

Estimated exponential annual growth rates of unrestrained populations range from low values of 0.077 for sheep in Tasmania and 0.091 for Northern elephant seals, to perhaps 1.0 for a pheasant population, 24 for the field vole, 10^{10} for flour beetles in laboratory cultures and 10^{30} for the water flea Daphnia. Human populations are at the lower end of this range, but a realistic exponential growth rate of 0.03 (or slightly above 3% per year) for some human populations is equivalent to a doubling time of about 23 years and a roughly thousand-fold increase in 230 years. Clearly, no population can maintain such a growth rate for long. (Expansion at the estimated annualized rates just cited for the field vole, flour beetle, and water flea is necessarily utterly fleeting).

The exponential growth rate of a population can be calculated from the schedule of fecundity at age x (b_x) and survival to age l) in a population. These "life table" variables are related to population growth rate by the Euler, or characteristic, equation, whose solution requires matrix methods. When the life table is unchanging for a long period, a population assumes a stable age distribution, which is also an intrinsic property of the life table, and a constant exponential rate of growth. Thus, assuming constant birth and death rates, the growth trajectory of a population may be projected into the future. However, because populations are finite and births and deaths are random events, the expected size of a population in the future has a statistical distribution that may include a finite probability of 0 individuals, that is, extinction. As a general rule, the probability of extinction decreases with increasing population size and increasing excess of births over deaths.

Population Regulation

Balancing the growth potential of all populations are various extrinsic environmental factors that act to slow population growth as the number of individuals increases. High population density depresses the resources of the environment, attracts predators, and, in some cases, results in stress-related reproductive failure or premature death. As population size increases, typically death rates of individuals increase, birth rates decrease, or both. The result is a slower growth rate and a changed, usually older, population. The predominant model used by animal ecologists to describe the relationship of population growth rate to population size (or density) is the logistic equation, in which the exponential growth rate of the population decreases linearly with increasing population size: where r_0 is the exponential growth rate of a population unrestrained by density (i.e., whose size is close to 0) and K represents the number of individuals that the environment can support at an equilibrium level, also referred to as the carrying capacity of the environment. Accordingly, the rate of growth of the population is expressed as when $N < K$, the growth rate is positive and the population grows. When $N > K$, the density-dependent term $(1 - N/K)$ is negative and the population declines. When $N = K$, the growth rate is 0 and a stable, steady-state population size is achieved. This depressing impact of density on the population growth rate is known as negative feedback.

The differential form of the logistic equation may be integrated to provide a function for the trajectory of population size over time. The curve is sigmoid (S-shaped), with the rate of growth, dN/dt reaching a maximum (the inflection point) at $N = K/2$. Because this is the density at which individuals are added to the population most rapidly, the inflection point also represents the size of the population from which human consumers can remove individuals at the highest rate without causing the population to decline. Thus, the inflection point is also known as the point of maximum sustainable yield.

Density dependence can take on a variety of forms. One of these is a saturation model where the exponential growth rate remains constant and positive until a population completely utilizes a nonrenewable resource such as space, and population growth stops abruptly. The approach of a population to an equilibrium level determined by density-dependent processes can be altered by environmentally induced changes in the intrinsic rate of population growth or in the carrying capacity of the environment.

Difficulties in finding mates and maintaining other social interactions at low densities, including group defense against predators, may also cause the population growth rate to decrease as density declines (the Allee effect), and, below a certain density threshold, may even result in population decline to extinction. This type of response is a positive feed-back, one that promotes population instability. For example, after commercial hunting had reduced populations of the passenger pigeon to low levels, the decline in social interactions in this communally nesting species is thought to have doomed it to extinction.

Populations have inherent oscillatory properties that can be triggered by time lags in the response to changing density and which cause populations to fluctuate in a perpetual limit cycle, with alternating population highs and lows. In these cases higher values of r can send a population into unpredictable chaotic behavior, increasing the risk of extinction. In a population with continuous reproduction, regular population cycles occur when there is a lag, often equal to the period of development, in the response of a population to its own density effects on the environment. When the time lag is of period τ, limit cycles develop when $r\tau$ exceeds $\pi/2$, and the period of the cycle is 4 to 5 times τ.

Metapopulations

Most natural populations consist of many subpopulations occupying patches of suitable habitat surrounded by unsuitable environments. Oceanic islands and freshwater ponds are obvious examples. But fragmentation of forest and other natural habitats resulting from clearing land for agriculture or urban development is increasingly creating fragmented populations in many other kinds of habitats. These subpopulations are connected by movement of individuals, and the set of subpopulations is referred to as a metapopulation. Metapopulations have their own dynamics determined by the probabilities of colonization and extinction of individual patches. A set of simple metapopulation models describes changes in the proportion of patches occupied (). When the extinction probability (e) of an individual patch is independent of p, the rate of loss of subpopulations is simply pe. The rate of colonization is proportional to the number of patches that can provide potential colonists and the proportion of empty patches that are available to receive them. Hence, colonization is equal to cp(1-p), where c is the rate of colonization.

The metapopulation achieves a steady state of number of patches occupied when colonization balances extinction, that is $pe = cp(1-p)$, or $\hat{p} = 1-e/c$. In this model, as long as the rate of colonization exceeds that of extinction, the metapopulation will persist. In more complex models, particularly when the probability of population extinction is reduced by continuing migration of individuals between patches (which keeps the sizes of subpopulations from dropping perilously low), the extinction rate and colonization rate both depend on patch occupancy. In this case, the solution to the metapopulation model has a critical ratio of colonization to extinction, below which patch occupancy declines until the metapopulation disappears. Thus, changes in patch size or migration between patches can cause an abrupt shift in the probability of metapopulation persistence.

Predator-Prey Interactions

The dynamics of populations are influenced by interactions with predator and consumer populations. Because these interactions have built-in lag times in population responses, they often result in complex dynamics. Among the most spectacular fluctuations in size are those in populations of snowshoe hares and the lynx that prey on them. Population highs and lows may differ by a factor of 1,000 over an oscillation period of about ten years. Oscillation periods in other population cycles of mammals and birds in boreal forest and tundra habitats may be either approximately four years or nine to ten years.

The biologists Alfred Lotka and Vito Volterra independently developed models for the cyclic behavior of predator-prey systems in the 1920s. The most basic model expresses the rate of increase in the prey population in terms of the intrinsic growth capacity of the prey population and removal of prey individuals by predators, which is proportional to the product of the predator and prey population sizes. The growth of the predator population is equal to its birth rate, which depends on how many preys are captured, minus a density-independent term for the death of predator individuals. The joint equilibrium of the prey and predator populations is determined by the predation efficiency and the relative rates of birth and death of the prey and predator, respectively. However, the equilibrium is neutral, which means that any perturbation will set the system into a persisting cycle. More complex models of predator-prey interactions include a balance between various stabilizing factors, such as density-dependent control of either population, alternative food resources for predators, and refuges from predators at low prey densities, and destabilizing factors, such as time lags in the response of the predator and prey to each other. For the most part, these models predict stable predator and prey populations under constant conditions.

Both empirical and experimental studies have shown that the rate of predation is nonlinear, violating one of the assumptions of the Lotka-Volterra model. When predation is inefficient at low prey densities and predator populations are limited by density-dependence at high predator densities, there may be two stable points. One of these is at a high prey population level limited by the prey population's own food supply, the other at a low prey population level limited by predators. When a prey population, such as a crop pest, is released from predator control following depression of the predator population by extrinsic factors such as climate, disease, pesticides, and so on, the prey may increase to outbreak levels and become a severe problem. Thus, agricultural practices that incidentally depress the populations of natural control organisms can have unwanted consequences.

A special kind of predator-prey model is required to describe the interactions between parasites, including disease-causing organisms, and their hosts. These models need to take into account the fact that parasites generally do not kill their hosts, that the spread of parasites among hosts may depend on population density and the presence of suitable vectors, and that hosts may raise defensive immune reactions. Immune reactions create a time lag in the responses of parasite and host populations to each other and may result in strong fluctuations in the prevalence of parasitic diseases.

Plant Ecology

Plants are one of the densest ecotypes with vicariate diversity in their life history depending on their mores. The ecology of land entirely depends upon plants as they are practical pioneers of the planet earth. Initiating from microorganisms to the macroorganism, they meet their living requisites expending plants. Admitting this, different vegetation types, such as grassland and timberland, are major biomes which help to flux the environmental shifts under global impact. The day-by-day advances in science and technology and ever-emerging needs of living organisms are some anthropogenic activities which had entirely changed the scenario of green systems of the world. Such drastic changes in plant ecology are harmful for all the beings and other respective factors. These substantial amendments occurred due to the earthquakes, industrial effluents, forest fires, carbon ignition, destruction of vegetative and agronomic landforms, etc.

Traditional Ecology

Broadly stating the relationship of plants with the other living organisms and environmental factors is called plant ecology. But elaborating the term, plant ecology is the study of the relationship of plants with the biotic (living organisms such as animals and other plants, bacteria, and fungi) and abiotic factors such as moisture, temperature, sunlight, soil (nutrients and salinity), and water (rainfall and water table) surrounding them. By the passage of time, the addressed issues regarding biosphere came into consideration. Though from the time of Alexander Von Humboldt (father of ecology) the known species of plants were about 20,000, now the number increased up to 40,0000 but the changes in elements of biosphere are increasing the issues such as loss of habitat, plant, animal, microflora, mutation, pollution, and soil sickness.

Due to these issues, the most affected living organisms are plants which urged the scientists to investigate the root causes of such drastic changes and commotion to the plant ecology. According to the climate, human and animal interaction, flora, and fauna, the planet earth is categorized into biomes. There are about six major biomes with cutting clarity of subcategories. The largest biome is boreal/coniferous forest; however, the second largest biome is grasslands that are ubiquitous as compared to other biomes; tropical rainforest covers only 6% of the world, but they have the richest biodiversity; however, the hottest biome is desert with the minimal biodiversity; in contrast the coldest biome is alpine forest merely with considerable biodiversity. Specifically, the plant populations have dominantly occupied this globe; according to estimation, 99.9% area of planet earth is covered with flora. About 350,000 species of plants excluding ferns, bryophytes, and algae are known and documented yet. Among them approximately 20% are under the risk of endangerment. The risk of endangerment or extinction due to natural and unnatural disasters has disturbed the whole food chain and web and is continually pushing toward the worst conditions.

Phylogenetic Ecology

Whenever ecological drift and loss in biodiversity of living organisms are discussed, it is generally apprehended that plants are vanishing due to overgrazing and animals are dying due to the inaccessibility of plants. But this whole globe is alive and functional on a single principle named balanced metabolic dynamics ratio between autotrophs (plants, producers) and heterotrophs (animals and microbes, consumers and decomposer). Basically, this trophic dynamics between producer and consumer within the biosphere is regulated by the transfer of energy from one part of the ecosystem to the other and even within the same ecosystem also known as energy flow in ecosystem. Except solar radiation (external source of energy), all the other energy systems are recycled and balance the dynamics of trophic level followed by complex metabolic mechanisms within the biosphere.

Drafting the origin of plant, their functional types and phylogenetic/evolutionary patterns are the most needed steps to timely track and record the drifts and risks to the ecosystem and biosphere. As the current dynamics, composition and distribution of plants are altered thence, elaborating and re-defining the relationship of plants with the factors encompassing them had led the flora and environment on the verge of endangerment are also expatiated, and many successful solutions to indemnify these issues are contributed by the scientists. Sustaining to this several concepts such as phylogeny, phenology, phytosociology, physiology, and anatomy of plants were used for modeling and surveying.

Terrestrial vegetation plays a phenomenal role in management of landscape and hydrological regime. Also the climatic change can be ameliorated by them as they could better regulate biogeological water cycle and sequestrate carbon cycle. The provision of protection against water resources by surface runoff leading toward flood attenuation, aquifer recharging, sea water leveling, water table leveling and fresh water management. Increase in temperature, variation in precipitation, and extreme events have potentially manifested the natural conversational and agricultural management regimes including an indirect risk that was constrained for social and human livelihoods. Although most of the commotion was inflicted due to water regime mismanagement, actually the fraction has been completed by burning practice, by grazing, or by harvesting hay/fodder.

Insect Ecology

Ecology is the study of the interrelationships between organisms and their environment. An insect's environment may be described by physical factors such as temperature, wind, humidity, light, and biological factors such as other members of the species, food sources, natural enemies, and competitors (organisms using the same space or food source). An understanding or at least an appreciation of these physical and biological (ecological) factors and how they relate to insect diversity, activity (timing of insect appearance or phenology), and abundance is critical for successful pest management.

Some insect species have a single generation per season (univoltine), while others may have several (multivoltine). The striped cucumber beetle, for example, overwinters as an adult, emerges in the spring, and lays eggs near the roots of young cucurbit plants. The eggs hatch, producing larvae that emerge as adults later in the summer. These adults overwinter to start the cycle again the next year. In contrast, egg parasitoids like Trichogramma overwinter as immature within the egg of their host. During the summer they may have several generations.

Insects adapt many types of environmental conditions during their seasonal cycle. To survive the harsh winters, cucumber beetles enter a dormant state. While in this dormant state, metabolic activity is minimal and no reproduction or growth occurs. Dormancy can also occur at other times of the year when conditions may be stressful for the insect.

It is often better to consider insects as populations rather than individuals, especially within the context of an agro-ecosystem. Populations have attributes such as density (number per unit area), age distribution (proportion in each life stage), and birth and death rates. Understanding the attributes of a pest population is important for good management. Knowing the age distribution of a pest population may indicate the potential for crop damage. For example, if most of the striped cucumber beetles are immature, direct damage to the above ground portions of the plant is unlikely. Similarly, if the density of a pest is known and can be related to the potential for damage, an action may be required to protect the crop. Information about death rates due to natural enemies can be very important. Natural enemies do nothing but reduce pest populations and understanding and quantifying their impact is important to effective pest management. This is all the more reason to conserve their numbers.

Components Factors of Environment

Following components of environment affect or influencing on insect population:

- Abiotic Factors: (Physical/non-living /Density independent factor).
    - Climatic factor:
        - Temperature.
        - Rainfall.
        - Humidity (Moisture).
        - Air current (Wind).
        - Light.
        - Atmospheric pressure.
    - Topographic factor:
        - Mountains.
        - Sea, Ocean, River.
        - Soil.
- Biotic factors: (Living /Density dependent factor):
    - Food (Nutritional factor).
    - Competition (Interspecific and Intraspecific Competition).
    - Natural enemies (Predators, Parasite and Pathogen).

Abiotic Factors (Density Independent Factors)

Temperature

- Insects are the cold blooded animals; they do not have mechanism to regulate body temperature called poikilothermic.
- Insects survive at specific optimum temperature - Upper lethal limit i s 40-50 °C (even up to 60 °C survival in some stored product insects) and Lower lethal limit - Below freezing point e.g. snow fleas.

- At low temperature (winter) insect takes more days to complete a stage (larval or pupal stage) Larva, pupa commonly undergoes hibernation in winter.
- At high temperature (summer) it takes less than to complete a stage. Eggs undergo aestivation in summer.
- Temperature effects on fecundity, migration and rate of development of insects.
- Example: Grasshopper lays 20-30 times more eggs at 32 °C compared to 22 °C, Oviposition of bed bug inhibited at 8-10 °C, Thrips give few eggs at 8 °C and more at 20-30 °C, Larval period of sugarcane internode borer isvery short in summer and prolonged in winter and Swarm migration oflocust occurs at 17-20 °C.

Rainfall

- Heavy rainfall has adverse effect on small insects like aphids, Jassids, thrips, white fly, mealy bugs, diamond back moth (DBM) and scale insects etc. which are washed out from plants and killed in flooded soil.
- Rainfall also effect on the abundance of insects.
- Rainfall is essential for adult emergence of cutworms and RHC.

Humidity/Moisture

- It is essential for physiological activities like metabolic reactions and transportation of salts in insects.
- Insect get die when water content increases or decreases termed as lethal wetness or lethal dryness. Moisture scarcity leads to dehydration and death of insects.
- High humidity causes development/Encourages of disease causing pathogens on insects (fungi). e. g. White halo fungus Verticillium lecaniioncoffee green scale requires high RH for multiplication and spread, Termites prefer high humidity 90-95% RH.
- Humidity also effect on fecundity and normal development of insects.
- Example: Locust does not lay eggs if there is no sufficient moisture. It mature quicker and give more eggs at 70% relative humidity, Low RH in rainfed groundnut crop induces leaf mines incidence.

Air current (Wind)

- The disposal of insects to great extent depends upon wind.
- It interferes with feeding, mating and oviposition of insects.
- Many insects fly with the air current and get transfer from one place to other. Many of them die by falling in rivers or sea. Many insects are known to spread in new countries through air currents.
- Large number of aphids has been found after a strong wind and many of them are destroyed by falling in the sea and rivers etc.
- Thus air current also play an important role in natural control of insects.

Light

- Light plays an important role in growth, development and survival because it is prime source of energy in all organisms.
- Light control locomotive activities of insects by direct action this phenomenon is called as Phytokinesis. The movement of animal in response to light called phototaxis. The response of organism towards the length of day light called as Photoperiodism.
- Photo period influences induction of diapause (a resting stage) in mostofthe insects E. g. Short day species-mulberry silk moth and Long Day Species-Pink bollworm.
- Some insects are active in night (Nocturnal). Some are active during the day (diurnal) and some active during dawn and dusk (Crepuscular).
- In many insects oviposition is stimulated by exposure to light or darkness. E.g. Fruit flies lays eggs in light, Lepidopterans like cotton bollworm, Red hairy caterpillar (RHC) oviposit in dark.

Atmospheric Pressure

- The insects are more affected in low atmospheric pressure than in high atmospheric pressure.
- The phototrophic insects are more active during periods of high atmospheric pressure.
- The activities of some insects are directly influenced by pressure. E.g. due to the low atmospheric pressure a chances of rain increased resulting the emergence of ants.

Topographic Factors

- Major topographic factors like mountain, rivers, sea are act as physical barriers to the spread of insects.
- Lake and ponds affect the nature of insects of that region.
- Water current: Larva of mosquitoes and beetles are able to live only in standing water and running water is preferred by Dragonfly and Caddis flies.
- Soil type: Wire worm, multiplies in heavy clay soil with poor drainage whereas termites, white grubs and cut worm prefer light, loamy soil.

Biotic Factors (Density Dependent Factors)

Food (Nutritional Factor)

- Insects are heterotrophic hence they cannot synthesize their own food they depend on plants for food.
- The quantity and quality of food/nutrition plays important role insurvival, longevity, distribution, reproduction and speed of development.
- Quantity of food: Short supply of food causes intraspecific and interspecific competition.
- Quantity of food: This depends on nutritional availability of plants. Crop varieties/species differ in nutritional status which affects insects.

Competition

Insect species are likely to be competingwithoneanotherorwith members of another speciesfor limited resources like food, mates, suitable site for oviposition or pupation. Such competition operates whenever the population is increasing and the resources are limited.

- Intraspecific competition: When members of population of the same species compete for resources. Examples are as follows:
    - Cannibalism in American bollworm larvae.
    - Cannibalism in later stage grubs of Chrysopid.
    - Crowding in aphids result in alate (winged) form for migration.
    - Reduction in fecundity (egg laying) in rice weevil during overcrowding.
    - Crowding in honeybees leads to swarming.
- Interspecific competition: This is the competition occurring between members of two or more species. Two or more competing species with identical requirements cannot coexist in a same place for a long time. The elimination of one species by another as a result of interspecificcompetition. It gives by Russian scientist G. F. Gause called competitive exclusion principle or Gause's hypothesis/principle.
- Example: Accidental introduction of oriental fruit fly into Hawaii eliminated by Mediterranean fruit fly, Trichogramma and Crysoperla compete for Helicoverpa eggs in cotton:
    - Cannibalism: It is the phenomenon where the insects feed on the individuals of the same species when crowding is occurring.
    - Colonization: Grouping of free living individuals to form colonies to have better protection from natural enemies or environmental conditions for improved utilization of food.
    - Aggregation: Tendency of congregating in large numbers than normal distribution for mating, food etc.

Natural Enemies

Every insect has a number of natural enemies in the nature viz; parasitic insects, predatory insects, mites, spider, birds, mammals, reptiles, fishes and diseases causing fungi, bacteria and viruses. They keep the insect population in check and thus natural balance within limits is almost always maintained.

Predators

Predators are free living organisms that feed on living insects and consume more than one individual during their lifespan. Major insect predators are as follows:

- Lady bird beetle (Coccinellids): Feed on aphids and leaf hoppers.
- Green lace wing (Chrysoperla spp.): Feed on aphids and other sucking pests.

- Mantid, Dragon fly and Syrphid fly: Feed on different insects.

Parasitoid

An insect parasite of an arthropod that is parasitic in its immature stage killing the host in the process of development andadults are free living. Major insect parasitoids are as follows:

- Trichogramma spp: Feed on Sugarcane borers and Cotton bollworms.
- Apanteles flavips: Feed on lepidopteran larvae.
- Epiricania meloneuca: Feed on sugarcane pyrilla.
- Capidosoma koeheleri and Chilonus blackberni: Feed on potato tuber moth.

Parasites

Parasites are organisms which live on other organisms for getting food and shelter. The association of parasite and host known as parasitism. A parasite weakens or kills the host while feeding requires only one part of one host to reach maturity. Example: Ticks, mites, Protozoa, Nematodes and other arthropods.

Pathogens

Disease causing microorganisms called as pathogens. Certain fungi bacteria and viruses cause disease in many insects and reduce their population in nature. The important microorganisms which cause disease in insects are as follows:

- Fungi: Example: Verticillium lecanii (White halo fungus) cause disease in mealybugs and aphids, Beaveria bassiana (White muscardine fungus) cause disease in lepidopteran larvae and Metarhizium anisopliae (Green muscardine fungus) in coconut rhinoceros beetle.
- Bacteria: Example: Bacillus thuringiensis effective against lepidopteran larvae & Bacillus popilliae attacking on beetles.
- Virus: Example: HaNPV (Helicoverpa armigera nuclear polyhedrosis viruses) against American bollworm and SlNPV - (Spodoptera litura nuclear polyhedrosis viruses) against tobacco leaf eating caterpillar.
- Other Natural Enemies: Frog, Toad, reptiles, birds, rats, bears, snakes and lizard.

Microbial Ecology

Microbes and Ecosystem Niches

Every ecosystem on Earth contains microorganisms that occupy unique niches based on their specific metabolic properties. Microbial life is amazingly diverse and microorganisms quite literally cover the planet. In fact, it has been estimated that there are 100,000,000 times more microbial cells on the planet than there are stars in the observable universe. Microbes live in all parts of the biosphere where there is liquid water, including soil, hot springs, the ocean floor, acid lakes, deserts, geysers, rocks, and even the mammalian gut.

By virtue of their omnipresence, microbes impact the entire biosphere; indeed, microbial metabolic processes (including nitrogen fixation, methane metabolism, and sulfur metabolism) collectively control global biogeochemical cycling. The ability of microbes to contribute substantially to the function of every ecosystem is a reflection their tremendous biological diversity.

Microbes are vital to every ecosystem on Earth and are particularly important in zones where light cannot approach (that is, where photosynthesis cannot be the basic means to collect energy). Microorganisms participate in a host of fundamental ecological processes including production, decomposition, and fixation. They can also have additional indirect effects on the ecosystem through symbiotic relationships with other organisms. In addition, microbial processes can be co-opted for biodegradation or bioremediation of domestic, agricultural, and industrial wastes, making the study of microbial ecology particularly important for biotechnological and environmental applications.

A Biofilm of Thermophilic Bacteria: Thermophiles, which thrive at relatively high temperatures, occupy a unique ecological niche.

Each species in an ecosystem is thought to occupy a separate, unique niche. The ecological niche of a microorganism describes how it responds to the distribution of resources and competing species, as well as the ways in which it alters those same factors in turn. In essence, the niche is a complex description of the ways in which a microbial species uses its environment.

The precise ecological niche of a microbe is primarily determined by the specific metabolic properties of that organism. For example, microbial organisms that can obtain energy from the oxidation of inorganic compounds (such as iron-reducing bacteria) will likely occupy a different niche from those that obtain energy from light (such as cyanobacteria). Even among photosynthetic bacteria, there are various species that contain different photosynthetic pigments (such as chlorophylls and carotenoids) that allow them to take advantage of different portions of the electromagnetic spectrum; therefore, even microbes with similar metabolic properties may inhabit unique niches.

Organization of Ecosystems

Microorganisms serve essential roles in the complex nutrient exchange system that defines an ecological community. Although ecologists tend to regard ecosystems as basic structural units, it can be difficult (if not impossible) to formally define the boundaries of a given ecosystem. As such, ecosystems are better thought of as conceptual rather than actual geographical locations. Rarely are ecosystems isolated from one another; rather, they should be considered parts of a larger functioning whole that together comprise the biosphere ("the place on Earth's surface where life dwells").

Despite the fact that clear boundaries between ecosystems may be difficult to identify, the myriad interactions that take place within an ecological community can often be observed and defined. These interactions may be best described by detailing feeding connections (what eats what) among biota in an ecosystem, thereby linking the ecosystem into a unified system of exchange.

All life forms in an ecosystem can be broadly grouped into one of two categories (called trophic levels):

- Autotrophs: Which produce organic matter (food) from inorganic substances?
- Heterotrophs: Which must feed on other organisms in order to obtain organic matter?

In general, trophic levels are used to describe the way in which a particular organism within an ecosystem gets its food. Using this description, we can restate and reorganize the categories above to define the three basic ways organisms acquire their food:

- Producers (autotrophs) do not usually eat other organisms but pull nutrients from the soil or the ocean and manufacture their own food using photosynthesis. In this way, it is the energy from the sun that usually powers the base of the food chain.
- Consumers (heterotrophs) cannot manufacture their own food and need to consume other organisms.
- Decomposers break down dead plant and animal material and wastes and release them into the ecosystem as energy and nutrients for recycling.

Within ecosystems, the biotic factors that comprise the categories above can be organized into a food chain in which autotrophic producers use materials and nutrients recycled by decomposers to make their own food; the producers are in turn eaten by heterotrophic consumers. In real world ecosystems, there are multiple food chains for most organisms (since most organisms eat more than one kind of food or are eaten by more than one type of predator). Additionally, the movement of mineral nutrients in the food chain is cyclic rather than linear. As a consequence, the intricate network of intersecting and overlapping food chains for an ecosystem is more commonly represented as a food web. A food web depicts a collection of heterotrophic consumers that network and cycle the flow of energy and nutrients from a productive base of self-feeding autotrophs.

Microorganisms play a vital role in every ecological community by serving both as producers and as decomposers. Although plants are the most common primary producers, autotrophic photosynthetic microbes (such as cyanobacteria and algae) can harness light energy to generate organic matter. Additionally, in zones where light cannot penetrate (and thus photosynthesis cannot be the basic means to produce energy), chemosynthetic microbes provide energy and carbon to the other organisms in the ecosystem. Other microbes are decomposers, with the ability to recycle nutrients from dead organic matter and other organisms' waste products. Decomposition is critical as most of the carbon and energy incorporated into plant tissues during photosynthesis remains uneaten when the plant tissue dies (and therefore must be broken down before it can be made available for recycling).

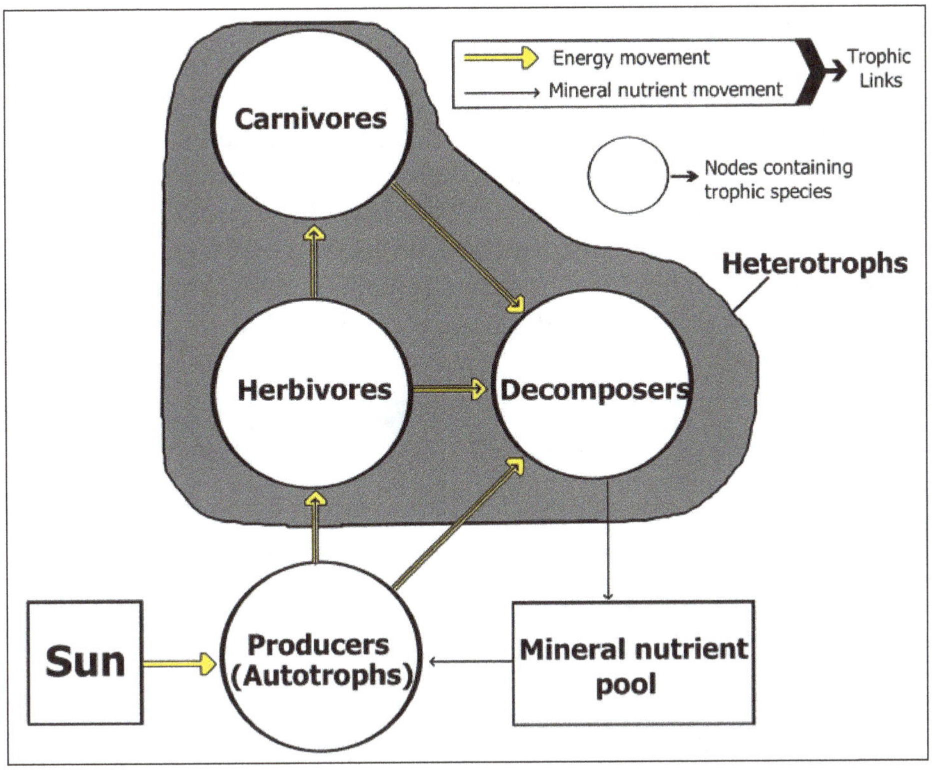

A simplified food web: A simplified food web model of energy and mineral nutrient movement in an ecosystem. Energy flow is unidirectional (noncyclic) and mineral nutrient movement is cyclic.

Role of Microbes in Biogeochemical Cycling

Microbes form the backbone of every ecological system by controlling global biogeochemical cycling of elements essential for life.

Microbial Role in Biogeochemical Cycling

Nutrients move through the ecosystem in biogeochemical cycles. A biogeochemical cycle is a pathway by which a chemical element (such as carbon or nitrogen) circulates through the biotic (living) and the abiotic (non-living) factors of an ecosystem. The elements that move through the factors of an ecosystem are not lost but are instead recycled or accumulated in places called reservoirs (or "sinks") where they can be held for a long period of time. Elements, chemical compounds, and other forms of matter are passed from one organism to another and from one part of the biosphere to another through these biogeochemical cycles.

Ecosystems have many biogeochemical cycles operating as a part of the system. A good example of a molecule that is cycled within an ecosystem is water, which is always recycled through the water cycle. Water undergoes evaporation, condensation, and then falls back to Earth as rain (or other forms of precipitation). This typifies the cycling that is observed for all of the principal elements of life.

Although biogeochemical cycles in a given ecosystem are coordinated by the full complement of living organisms and abiotic factors that make up that system, microorganisms play a primary role in regulating biogeochemical systems in virtually all of our planet's environments. This includes extreme environments such as acid lakes and hydrothermal vents, and even includes living systems such as the human gut. The key collective metabolic processes of microbes (including nitrogen fixation, carbon fixation, methane metabolism, and sulfur metabolism) effectively control global biogeochemical cycling. Incredibly, production by microbes is so immense that global biogeochemistry would likely not change even if eukaryotic life were totally absent. Microbes comprise the backbone of every ecological system, particularly those in which there is no light (i.e. systems in which energy cannot be collected through photosynthesis). Two key examples of critical biogeochemical processes carried out by microorganisms are discussed below.

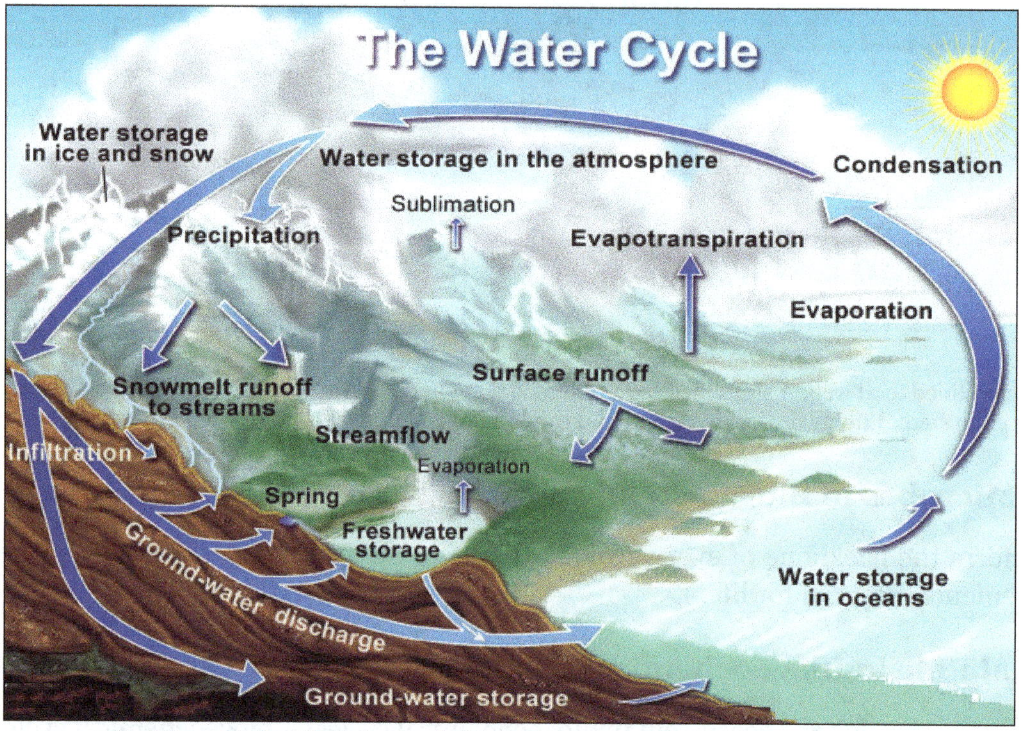

The Water Cycle: Water is recycled in an ecosystem through the water cycle.

The Carbon Cycle

Carbon is critical for life because it is the essential building block of all organic compounds. Plants and animals utilize carbon to produce carbohydrates, fats, and proteins, which can then be used to build their internal structures or to obtain energy.

Carbon in the form of carbon dioxide (CO_2) is readily obtained from the atmosphere, but before it can be incorporated into living organisms it must be transformed into a usable organic form. The transformative process by which carbon dioxide is taken up from the atmospheric reservoir and "fixed" into organic substances is called carbon fixation. Perhaps the best known example of carbon fixation is photosynthesis, a process by which energy derived from sunlight is harnessed to form organic compounds. Photosynthesis depends on the activity of microorganisms such as cyanobacteria; indeed, the fact that there is oxygen in the Earth's atmosphere at all is a consequence of the photosynthetic activity of ancient microbes.

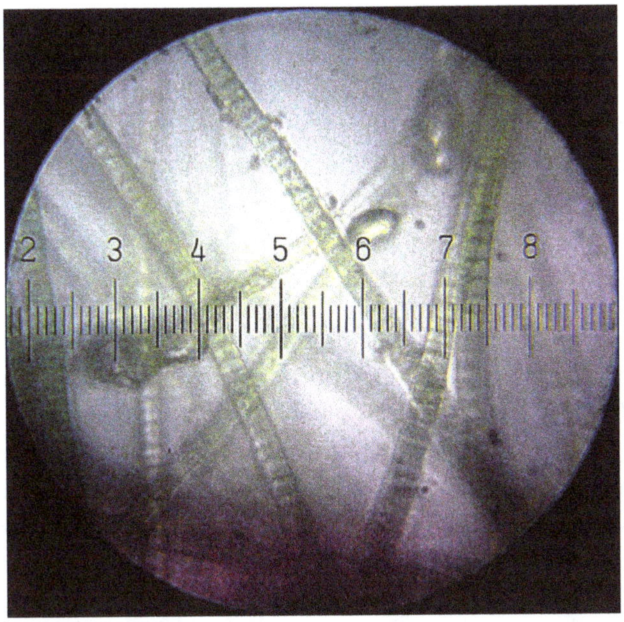

Cyanobacteria: Cyanobacteria, also known as blue-green bacteria, blue-green algae, and Cyanophyta, is a phylum of bacteria that obtain their energy through photosynthesis.

The Nitrogen Cycle

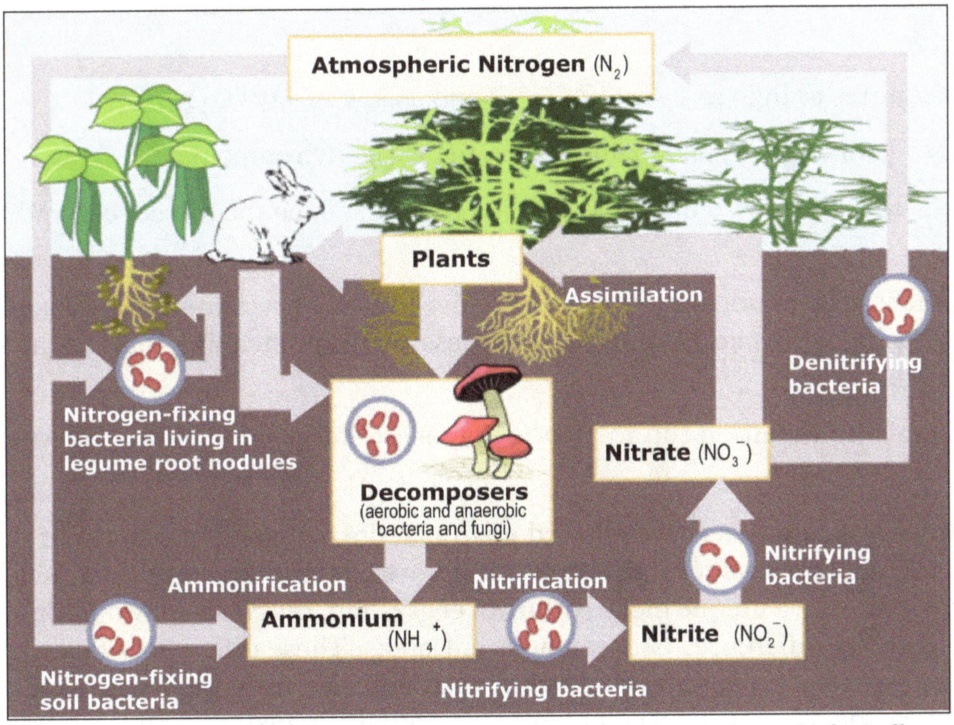

The Role of Microbes in the Nitrogen Cycle: The processing of nitrogen into a biologically useful form requires the activity of microorganisms.

Nitrogen is essential for all forms of life because it is required for synthesis of the basic building blocks of life (e.g., DNA, RNA, and amino acids). The Earth's atmosphere is primarily composed of nitrogen, but atmospheric nitrogen (N_2) is relatively unusable for biological organisms. Consequently, chemical processing of nitrogen (or nitrogen fixation) is necessary to convert gaseous nitrogen into forms that living organisms can use. Almost all of the nitrogen fixation that occurs on the planet is carried out by bacteria that have the enzyme nitrogenase, which combines N_2 with hydrogen to produce a useful form of nitrogen (such as ammonia). Thus, microorganisms are absolutely essential for plant and animal life forms, which cannot fix nitrogen on their own.

Microbial Environments and Microenvironments

The extraordinary biological diversity among microbes reflects their ability to occupy every habitable environment on the planet. Microorganisms are found on practically every habitable square inch of the planet. They live and thrive in all parts of the biosphere where there is liquid water, including hostile environments such as the poles, deserts, geysers, rocks, and the deep sea. Additionally, while microbes are often free-living, many have intimate symbiotic relationships with other larger organisms. Clearly, microbes have adapted to extreme and intolerant conditions, and it is this adaptation that has yielded tremendous biological diversity among microorganisms.

Like all extant organisms, microbes have evolved to thrive within a given environmental context. Microorganisms are ubiquitous despite the fact that the planet is host to extraordinarily diverse environments. Therefore, microbes have adapted to fill every ecological niche on the planet. For example, extremophilic species have been found that can tolerate the following environmental extremes:

- Temperatures as high as 130 °C (266 °F) and as low as −17 °C (1 °F).
- Highly alkaline (pH 0) and highly acidic (pH 11.5) environments.
- Extremely saline environments (including those in which the salt concentration is saturating).
- Extremely high (1,000-2,000 atm) and low (0 atm) pressures (some bacteria can survive for prolonged periods in a pressure-less vacuum, meaning they might even survive in space).
- High ionizing radiation (up to 15,000 Gy; as a reference, a mere 5 Gy would kill a human).

These evolutionary adaptations have allowed microbial life to extend into much of the Earth's atmosphere, crust, and hydrosphere (the water found over, under, and on the surface of a planet). In addition to occupying a unique niche within an ecosystem, microbes are potentially sensitive to subtle environmental differences between adjacent areas. These differences define so-called microenvironments (or microhabitats) that can be distinguished from the immediate surroundings by such factors as the amount of incident light, the degree of moisture, and the range of temperatures. For example, the side of a tree that is shaded from sunlight is a microenvironment that typically supports a somewhat different community of microorganisms than would be found on the side that receives regular light. Microbes, therefore, are not only adapted to their habitat, but also to the immediate environment, thus promoting increased diversity among microbial species within an ecosystem.

Ecology by Biome

Forest Ecology

Forest ecology is the study of forest ecosystems. Forests are ecosystems in which the major ecological characteristics reflect the dominance of ecosystem conditions and processes by trees. Ecosystems are ecological systems that have the attributes of structure, function, interaction of the component parts, complexity (that reflects the structure, function and interactions) and change over time. An ecosystem can be of almost any physical size as long as it exhibits these key characteristics, from a single plant growing in soil, to the entire world ecosystem.

The key structural components of forest ecosystems are plants, animals, microbes, soils and the atmosphere. Topography and microclimate are also important ecosystem features, but are not structural elements in the strict sense. The key functional aspects of forest ecosystems are energy capture and biomass creation; nutrient cycling and the regulation of atmospheric and water chemistry; and important contributions to the regulation of the water cycle. The interactions within an ecosystem involve all combinations of plant, animal and microbial interactions, interactions between organisms and the soil, and between the atmosphere and both the biotic community and the soil.

Complexity is an important attribute even though normally functioning forest ecosystems can exist at widely different levels of complexity. The importance of complexity lies in its implications for our ability to understand and predict, and therefore manage, forest ecosystems.

Forest ecosystems are continually changing. This change, initiated by external disturbance factors but largely determined by internal ecosystem processes, is vital for the maintenance of many aspects of biological diversity. In many types of forests it is essential for the long-term sustainability of the ecosystem.

"Forest stewardship" and "good, sustainable forestry" can only be defined in terms of society's desires and preferences with respect to stand and landscape-level forest conditions, functions and values. However, unless forestry is based on a respect for forest ecology and the ecological characteristics of forest ecosystems, it is very unlikely that society's long-term desires will be satisfied. Because of the long time scales of forestry, decisions about forest management must be founded on ecologically-based forecasts of ecosystem response, involving the use of ecosystem management simulation models.

Forests are local or regional segments of landscapes in which biological and ecological conditions and processes are dominated by the presence of trees - large, generally long-lived perennial plants characterized by a large woody stem and a large woody root system. The size and longevity of trees confer on them the ability to dominate other plant types by expropriating light and soil resources. This enables trees to control the major ecological processes, to determine the habitat for animals, microbes and other plant types, and to play a major role in determining the abundance of these other organisms in the forest. Trees also dominate the hydrological cycle, the soil development processes, the microclimate and the ecological characteristics of streams in forest ecosystems. Forests can also be dominated by large plants with woody stems that are not strictly trees, such as bamboo or tree ferns.

Ecology is the science that studies ecosystems. An ecosystem is any ecological system that exhibits five key attributes:

- Structure: There are living and dead plants, animals and microbes arranged in vertical and horizontal patterns in local ecosystems and across regional landscapes. There is a physical environment that in terrestrial ecosystems consists of soil, geological substrates, and an atmosphere. In aquatic ecosystems, the medium is water, and the ecological processes may or may not be affected by sediments or geological substrates.
- Function: The system of living and dead organisms, together with their abiotic environment, results in the combining of physical (light) or chemical energy with chemical elements (nutrients) from the soil, geological substrate, water or atmosphere to create the complex, high-energy, organic molecules that make up living organisms and render life as we know it possible. The energy for the creation of these molecules in forest ecosystems is provided almost entirely by photosynthesis. The chemical raw materials for their synthesis are provided by the circulation of appropriate mixtures of different nutrient elements in the ecosystem (nutrient cycling), which in turn is strongly influenced by temperature and moisture, and hence by climate, and by the action of soil animals and microbes.
- Complexity: Forest ecosystems are characterized by complex assemblages of different plant life forms, from lichens and bryophytes (e.g., mosses) to various types of herbs and shrubs, to climbers and trees. Associated with these different living plants is an assemblage of animals and microbes that use them as a source of energy and nutrition. Dead plant and animal matter provides the energy resource for a different set of animals and microbes that serve to decompose the organic matter and make the nutrients they contain available again for uptake by a new generation of living plants and microbes. The interactions between the living organisms and between these organisms and both the dead organic matter and the physical environment contribute to the complexity of ecological systems.
- Interactions between ecosystem components: As is the case for all systems, a key feature of an ecosystem is the interaction between the structural components. For example, soils affect plants, but plants also influence soil development. Climate determines soil and vegetation development, but plants modify local climate to produce a microclimate; vegetation can sometimes influence regional climate features and global vegetation plays a major role in global climate. Animals and microbes affect plants, but plants largely determine which animals and microbes will be affecting them.
- Change over time: For humans, "nothing is as certain as death and taxes". For ecosystems, nothing is as certain as change. Just as individual animals are born, grow, mature and inevitably die, ecosystems undergo "renewal" as a result of ecosystem disturbance caused by humans or non-human processes or events. Ecosystems develop and "mature", and are eventually disturbed and the process of renewal is repeated. This change over time is as fundamental to ecosystems as any of their other characteristics, and in many cases is an essential requirement for their long-term stability and the maintenance of historical ranges of variation in their structure, function, complexity (including biological diversity) and the interactions of their components.

A forest ecosystem is an area of the landscape, varying in size from a local stand (a few hectares or less) to an entire continent, in which the structure, function, complexity, interactions and patterns of change over time are dominated by trees. Forest ecology is the study of these tree-dominated landscape units. Stand-level forest ecosystems are terrestrial ecosystems, but landscape-level forest ecosystems frequently include streams, rivers and lakes, and areas of non-forested terrestrial ecosystems. However, the overall character of these other types of ecosystems is strongly influenced by their location in a tree-dominated landscape.

Forests Different from one Part of the World to Another

The type of forest that can develop in an area is ultimately determined by climate. The broad bands of different vegetation types one can observe while traveling from sea level to the top of a tall mountain at the equator, or from sea level at the equator to sea level in the arctic, reflect changes in temperature and precipitation along these elevation/latitudinal gradients. Similarly, climatic gradients from the coast to the interior of continents, and with increasing distance away from large lakes, are associated with differences in vegetation and in the accompanying soils, animals and microbial communities. Tropical rain forest, tropical seasonal forest, tropical dry forest, subtropical forest, savanna, grasslands, evergreen warm and evergreen cool temperate forest, deciduous temperate forest, boreal forest, shrub lands and tundra are major vegetation types, each of which is associated with a particular climate that determines the potential vegetation that can develop, and the associated animal and microbial communities.

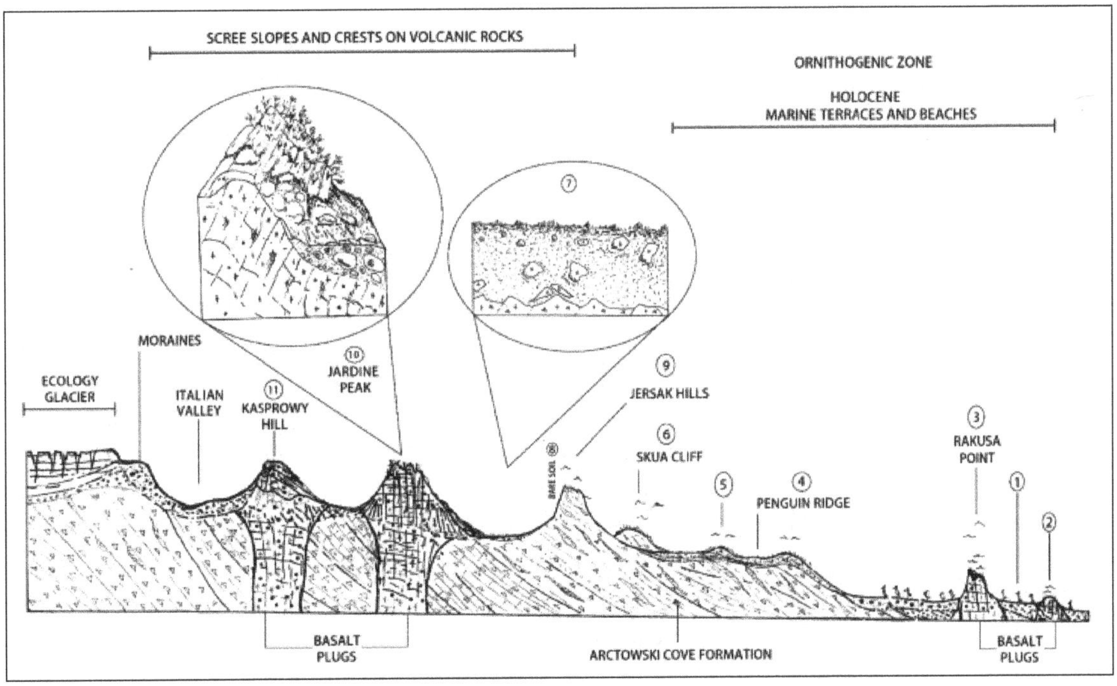

Variation in the structure and species composition of the plant community along a local gradient of variation in soil moisture and fertility that is associated with a local topographic gradient from the top of a small ridge to the valley below. The example is from a low elevation coastal forest.

Within a particular regional climatic area, the achievement of the climate-related vegetation potential is determined by aspect (which modifies the regional climate), the slope and the position of the ecosystem on the slope (which affects soils and soil moisture), and the soil and bedrock geology (which affects soil fertility and soil moisture). This ecological diversity (the diversity of climate, topography, geology and soil) creates both continental bands of broadly similar types of ecosystem (biomes), and local mosaics of biotic communities distributed along gradients of soil moisture and fertility, and across topographically-induced local variations in climate within the biomes.

Biomes - continental-level subdivisions of the continental ecosystem - consist of life zones (defined by major climatic belts and major topographic and geological features) and plant formations (plant communities dominated by plants of a particular life form; life forms include mosses, herbs, shrubs, deciduous trees, evergreen trees, etc.). Plant formations are in turn associated with particular groupings of animals (e.g., grassland animals, tundra animals, evergreen coniferous forest animals, deciduous forest animals) and microbes.

Forest biomes are a useful level of forest classification for continental and global-level studies of major differences between forests, and of relationships between forests and global climate, but biomes are too coarse a level of subdivision for many purposes. Understanding wildlife habitat relationships, biodiversity issues, sustainable forest management and forest stewardship requires an understanding of the local mosaics of forest ecosystem types within a biome. Because biomes are based on vegetation life form, the first subdivision of forest biomes is based on the dominant tree species in mature or old forests growing on sites of average moisture and soil fertility. This defines forested ecozones or biogeoclimatic zones.

Forest biogeoclimatic zones or ecozones are a mosaic of different forest ecosystem types growing on soils of different moisture and nutrient status within a particular climatic regime or area. These different ecosystem types vary in species composition, productivity, biological diversity and response to ecosystem disturbance. They also vary in the rates and patterns of change following disturbance.

Forests Change Over Time

Any gardener knows that vegetation changes if left to its own devices. Within a few years of neglect, most gardens will be taken over by a variety of "weeds" - unwanted native or introduced herbs and shrubs - and within a few decades most abandoned gardens in forested areas will be colonized by young trees. It is much less obvious that mature forests change. Once a long-lived tree is large, most people fail to notice as it gets taller and develops a thicker stem. Because most forests grow and change on time scales that are a significant fraction of, or longer than, a human lifetime, many people regard unharvest forests as a permanent ecosystem condition. In reality, change is always occurring, whether this is simply because branches are dying, because individual trees are getting larger, because natural processes are causing mortality that reduces the number of living trees in an area, or because fire, wind, snow, insects, disease epidemics or forest harvesting are periodically causing major modifications to ecosystem structure, species composition and function.

Forests change over time periods of millennia or longer due to climate change. They change more frequently - perhaps every one or few centuries - because of external physical factors (such as fire, wind, snow and landslides, referred to as allogeneic factors) or "external" biological factors (insect and disease epidemics, referred to as biogenic factors). In some forests, these external allogenic or biogenic disturbances may occur every few decades. Forests also change because of alterations in soil and microclimate, and in plant, animal and microbial species caused by within-ecosystem ecological processes at the population and biotic community level. This self-driven or autogenic change occurs following disturbance to the existing ecological condition caused by allogenic or biogenic disturbance processes. It is a process of ecosystem recovery back to, or towards, the pre-disturbance ecosystem condition, and is called ecological succession. Disturbance is defined as the action of allogenic and/or biogenic factors that results in rates, patterns and directions of successional change that are different from what would be expected in the ecosystem in question as a result of autogenic processes.

Changes in local and regional forest ecosystems over time are the combined result of all these processes: climate change, allogenic and biogenic change, and autogenic change. The details of the change - the degree and spatial scale of ecosystem disturbance, the rate of ecosystem recovery and the sequence of biotic communities that successively occupy and replace each other in areas that have been disturbed - are highly variable. Within any climatic area, there is an overall climatic control of the patterns and rates of autogenic recovery, but in many local ecosystems there can be a considerable range in the details, albeit within this overall control. Different temporal sequences of plant communities with different animal and microbial assemblages can occur according to the pre-disturbance forest condition, the nature and extent of the disturbance, the type of soil, and whether or not seeds or spores of particular plant species arrive in the disturbed area in time to participate in the developing plant community.

Biological diversity has become a topic of major concern around the world. The product of millions of years of evolution, biological diversity (or simply "biodiversity") is nature's "insurance policy" providing for ecosystem resilience in the face of climatic change and disturbance. It provides a vast wealth of economic and human health-related resources, and is a key part of the aesthetic, intellectual and spiritual forest values held by human society. Most people feel that the biodiversity of forests needs to be "protected" and "preserved" by minimizing ecosystem disturbance. However, the issue of biodiversity conservation is complex and requires an understanding of the origins and causes of biodiversity. These lie in:

- Ecological diversity, most of which is unaltered and unaffected by human activity.
- Ecosystem disturbance, which creates different, ages of forest with a wide variety of different structures and plant species composition, thereby providing habitat for a wide variety of animals and microbes. Ecosystem disturbance initiates new sequences of ecological succession that create temporal diversity in the forest.
- Autogenic ecosystem recovery (referred to as succession) which provides ever-changing soil, microclimatic and biotic conditions, ensuring that a wide variety of different species occur in any particular local ecosystem over time, and that there is a shifting mosaic of ecosystem conditions across regional landscapes that ensures a wide diversity of plant, animal and microbial habitats.

Conservation of biodiversity is thus related to appropriate types, frequencies, severities and scales of "natural" and human-caused disturbance (humans are also natural, as is the disturbance they cause), interacting with existing ecological diversity. Biodiversity objectives will only be achieved if biodiversity conservation plans are based on knowledge of how and why forest ecosystems vary from place to place, and how and why they change over time. Temporal diversity, the variation in all other measures of biodiversity over time, turns out to be the key to the maintenance of most other measures of biodiversity over longer time periods.

Types and Characteristic Feature

Temperate Forest Ecosystem

The temperate forest ecosystem is very important on Earth. Temperate forests are in regions where the climate changes a lot from summer to winter. Tropical rain forests are in regions where the climate stays constant all year long. Temperate forests are almost always made of two types of trees, deciduous and evergreen. Deciduous trees are trees that lose their leaves in the winter.

Evergreens are trees that keep them all year long, like pine trees. Forests can either be one or the other, or a combination of both. A fourth kind of forest is a temperate rain forest. These are found in California, Oregon and Washington in the United States. These forests are made of redwoods and sequoias, the tallest trees in the world. The amount of rainfall in an area determines if a forest is present. If there is enough rain to support trees, then a forest will usually develop. Otherwise, the region will become grasslands.

The Tropical Rainforest Ecosystem

Tropical rain forests are one of the most important areas on Earth. These special ecosystems are homes to thousands of species animals and plants. Contrary to popular belief, rain forests are not only densely packed plants, but are also full of tall trees that form a ceiling from the Sun above. This ceiling keeps smaller plants from growing. Areas where sunlight can reach the surface are full of interesting plants.

Do you know where rain forests get their name? They are so named because they receive a lot of rain – an average of 80 inches a year. The temperature doesn't change very much during the year. It is always warm and muggy. The famous Amazon jungle is located in Brazil, in South America. This particular forest is called the Neotropics. Other large blocks are located in Central and West Africa.

Insects of the Tropical Rainforest

The most feared and well known spider in the world resides in the jungle. Tarantulas are one of the creepiest animals you will ever see. Most species of tarantula have poisonous fangs for killing prey and for protection. Although some are life-threatening to humans, others are harmless. Army ants are just one species of ant in the rain forest. They are called army ants because they march in a long, thick line through the jungle. They only stop when the young larvae reach pupil stage. Once the queen lays its eggs, the ants start marching again.

Beautiful butterflies fill the forest, but at one time these insects weren't so pretty. Butterflies start out as caterpillars, which tend to be a tad on the ugly side. They go through metamorphosis, which is the process of changing into a butterfly. Centipedes aren't so lucky. They don't turn into butterflies, but instead roam the forest looking for food. Some centipedes use poison to kill their prey.

Tropical Rainforest Birds

The birds of the rain forest are the most beautiful in the world. A wide range of colors can be seen darting through the trees as the forest tops come to life. Many species of tropical birds are kept as pets because of their looks. Hundreds of species of parrot live in the rain forest. The scarlet macaw is just one of these. It is also one of the longest, stretching to a length of 3 feet from its head to the tip of its tail. When these macaws eat a poisonous fruit, they eat a special type of clay that neutralizes the poison.

Toucans are also very interesting birds. They have large beaks that they use to reach fruit they can't get to. Scientists estimate there are 33 species of toucan in the rain forest. Not every tropical bird was blessed with looks. The hoatzin looks more like a peacock without the pretty tail.

Hoatzins are terrible flyers – crash landings are common practice. The brown kiwi is a flightless bird that looks more like a rodent with a long beak and feathers. Kiwis live on the ground instead of the trees. They have special claws used for running, digging and defense.

Tropical Rainforest Mammals

Birds aren't the only creatures that fly through the rain forests. Several species of flying mammals live in the jungle. From the harmless fruit bat to the unique flying squirrel, the tropical rain forests are full of surprises. The flying fox is one of the largest bats in the world. Its wings can spread out to 5 feet in width. Unlike bats in other parts of the world, these bats do not live in caves. They prefer to .hang in trees during the day. Hundreds or even thousands of bats can be spotted in a single tree.

Vampire bats live in the Amazon jungle in South America. The famous stories of blood-sucking bats probably originated here. These bats do in fact drink the blood of their victims. They usually attack farm animals, but have also enjoyed the blood of humans. But vampire bats only drink a very small amount of fluid.

Tropical Rainforest Reptiles

The tropical rain forests of the world are full of reptiles. Reptiles are cold blooded, which means their body temperature depends on their environment. So, it is important for them to stay in warm climates. Snakes are reptiles, and the rain forests are home to many. The mamba family is the most poisonous of all. They kill their prey by injecting poison with their sharp fangs.

Anacondas make up another snake family. They are some of the longest creatures in the world, as they can reach 30 feet in length. Anacondas prefer to wrap themselves around their prey and squeeze, rather than inject poison. Anacondas swallow their prey whole and sleep while the food is digesting. Chameleons are interesting lizards that can change color.

This enables them to blend in with their surroundings. Not only is this a great disguise from predators, it is also an easy way to sneak up on their prey. Chameleons only eat insects. Geckos are very neat creatures. The flying gecko can glide from tree to tree to escape from predators. Their grip is so strong, that if you tried to pull one off a window, the glass would break before the gecko would let go.

Tropical Rain Forest Primates

Monkeys and their cousins are all primates. Humans are also primates. There are many species of monkeys in the tropical rain forests of the world. Monkeys can be divided into two groups: new world monkeys and old world monkeys. New world monkeys live only in South and Central America. Spider monkeys live in the rain forests in the Andes Mountains.

They look very strange with their long noses. Spider monkeys eat mostly fruit and nuts, so they are called frugivores. They are joined by the howler monkeys. These primates are so named because they have a special sac that makes their sounds louder.

Old world monkeys live only in Africa and Asia. The colobus monkey is one such kind. These monkeys are called foliovores because they eat leaves. They live in small groups of 15, but other primates live in larger groups of up to 200. There are too many species. Chimpanzees, orangutans and gorillas are all called pongids. These primates are more famous than the others. Gorillas are too big to climb trees, so they are found on the forest floor.

Boreal or Taiga Forests

The boreal forest ecosystem is the contiguous green belt of conifer and deciduous trees that encircles a large portion of the Northern Hemisphere. In North America, the boreal forest stretches across most of northern Canada and into Alaska. It has long been identified as one of the world's great forest ecosystems. This forest ecosystem covers roughly 35% of Canada's land mass and is the single largest land based ecosystem in North America. It also contains a significant proportion of Canada's biodiversity and has long been recognized as an important global carbon sink.

Although the boreal is relatively unknown, it is important as the "great lung" of North America, "breathing in" carbon dioxide and "exhaling" oxygen into the atmosphere. In short, the boreal forest manages to do what the rain forest of the Amazon does but with only the fraction of the flora and fauna.

This forest ecosystem houses the largest and smallest mammal species (wood bison and pygmy shrews) of the North American continent. The Boreal forest has many things: great lakes and northern rivers; vast bogs, fens and other organic wetlands. The rich wildlife diversity of the Boreal is a joy to behold: woodland caribou and lynx; whooping cranes and wood bison; northern owls; woodpeckers with three rather than four toes; colorful wood warblers.

The Boreal has more than 5,000 species of conspicuous and colorful fungi, celebrated far more in Scandinavia and Siberia than in North America. Then there are the precious old-growth forests, the richest and most biologically diverse of the Boreal forest communities that are essential for so many Boreal species.

Structure of Forest Ecosystems

Different organisms exist within the forest layers. These organisms interact with each other and their surroundings. Each organism has a role or niche in sustaining the ecosystem. Some provide food for other organisms; others provide shelter or control populations through predation:

- Producers: All living organisms' intake energy in order to survive. In a forest ecosystem, trees and other plants get their energy from sunlight. Plants produce their own food, in the form of carbohydrates. Plants are, therefore, called the primary producers, since they produce the basic foodstuffs for other organisms within food chains and food webs. Photosynthesis is the chemical reaction that allows plants to produce their own food.

- Consumers: Animals cannot produce their own food. They must consume food sources for die energy they need to survive. All animals, including mammals, insects, and birds, are called consumers. Consumers rely on plants and other animals as a food source. Primary consumers only eat plants and are referred to as herbivores. Secondary consumers are referred to as carnivores and feed on herbivores. Tertiary consumers are carnivores that feed on other carnivores. Omnivores eat both plant and animal matter.
- Decomposers: Leaves, needles, and old branches fall to the forest floor as trees grow. Eventually all plants and animals die. So what happens to all of this plant and animal material? No. Does it sit on the forest floor forever? These materials are decomposed by worms, microbes, fungi, ants, and other bugs. Decomposers break these items down into their smallest primary elements to be used again. Decomposers are important in that they sustain the nutrient cycle of ecosystems.
- Humans are part of Forest Ecosystem: Humans are consumers. We get food and materials from forests. Because of this, we are a part of the forest ecosystem. Human consumption alters forest ecosystems. Human intervention may be necessary to sustain forest communities under the increased pressure of human use.

Marine Ecology

Marine Ecology is the scientific study of marine-life habitat, populations, and interactions among organisms and the surrounding environment including their abiotic (non-living physical and chemical factors that affect the ability of organisms to survive and reproduce) and biotic factors (living things or the materials that directly or indirectly affect an organism in its environment). Marine ecology is a subset of the study of marine biology and includes observations at the biochemical, cellular, individual, and community levels as well as the study of marine ecosystems and the biosphere.

The study of marine ecology also includes the influence of geology, geography, meteorology, pedology, chemistry, and physics on marine environments. The impact of human activity such as medical research, development, agriculture, fisheries, and forestry is also studied under marine ecology. In some ways, marine ecology is more complex than the relatively straightforward study of a particular organism or environment because of the numerous interconnections, symbiotic relationships, and influence of many factors on a particular environment.

To understand the difference between marine biology and marine ecology, it may be useful to look at a community of organisms. A marine biologist may focus on behavioral relationships between the organisms in one particular species while someone studying ecology would study how the behavior of one organism influences another. An ecologist would also look at abiotic factors and how they influence that organism. A scientist studying community ecology might study a group of organisms to see how they influence other species and abiotic factors. The major subcategories of ecology are:

- Physiological ecology: The study of how biotic and abiotic factors act on the physiological characteristics of an organism and how the organism adapts to the abiotic and biotic environment.
- Behavioral ecology: A subcategory of ecology that studies which ecological and evolutionary dynamics is responsible for the way in which organisms adapt to their environment.
- Population ecology: The study of populations of organisms in a particular species and how the populations interact with their environment.
- Community ecology: The study of how species react to each other in a community.
- Landscape ecology: The study of how organisms interact with a particular landscape.
- Ecosystem ecology: The study of how energy and matter flow through ecosystems.
- Global ecology: The study of how energy and matter interact in the entire web of life on Earth.

The study of ecology in general includes all of the subcategories listed above as they apply to marine ecology, animal ecology, plant ecology, insect ecology, arctic ecology, tropical ecology and desert ecology.

Sustaining Life on Earth

Earth has been divided by ecologists into four areas: the hydrosphere, the lithosphere, the atmosphere, and the biosphere. The hydrosphere refers to water on the planet, the lithosphere consists of soil and rocks, the atmosphere is the air, and the biosphere refers to all of the life on Earth. The biosphere can be visualized as a thin surface layer on the Earth from 11,000 m below sea level to 15,000 m above sea level, even though there are no permanent residents living in the atmosphere.

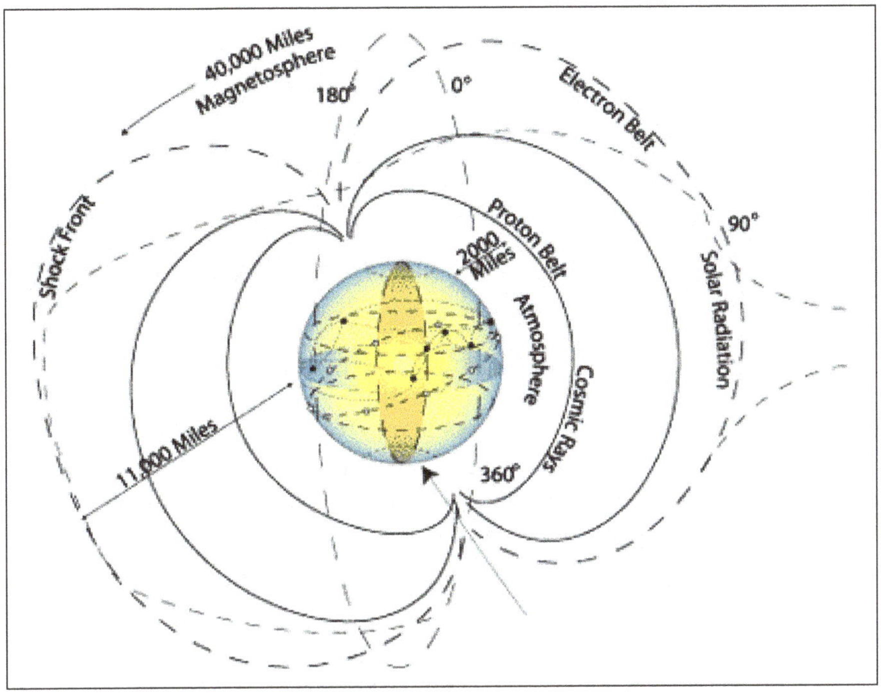

The Biosphere.

The first life on Earth was formed in the photic zone of the hydrosphere when organisms with more than one cell evolved in the deep ocean benthic zones. After the ozone layer formed, which protects land organisms from harmful UV rays, life began to evolve on land. After the continents separated and reformed, biodiversity began to increase as life began to adapt to new environments. Biodiversity can be observed at the genetic level, the species level, the population level, and the ecological level.

Abiotic elements like carbon, nitrogen and oxygen are present in great quantities in the biosphere. Phosphorus, calcium, and potassium are also present in smaller amounts. All are elements critical to the existence of life. Every element in the ecosystem transforms from mineral to organic forms and back to minerals and is never destroyed. Life depends on energy from the sun and the organisms that are capable of transforming light into chemical energy form the basis for the food chain. The process of photosynthesis converts light into chemical energy, resulting in the production of glucose and oxygen. Other organisms depend on glucose produced by photosynthesis for energy to fuel biological processes such as cellular respiration. During cellular respiration, organisms split glucose back into water and carbon dioxide. The breaking down of glucose releases and utilizes energy from the sun stored by photosynthesizing plants. The oxygen level of the Earth's atmosphere is largely reliant on the amount of photosynthetic activity and respiration going on in the biosphere. A buildup of elements in areas with a lot of organisms is prevented by circulation of the atmosphere with global air currents.

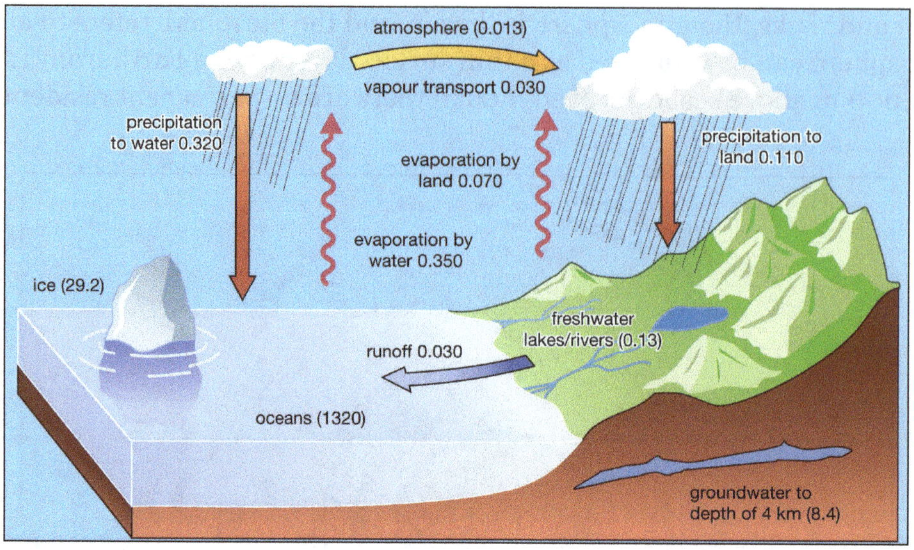

The Earth is full of cycles simultaneously occurring and interlaced. Water is cycled through the hydrosphere, lithosphere, atmosphere, and biosphere in relatively predictable movements. The flow of basic elements and the stability of the Earth's climate and temperature depend on oceanic currents and the vast amount of water stored in the ocean. Ecologists often depend on computer modeling to determine how human activity can influence the intricate cycling of the biosphere.

Every organism is influenced somehow by every part of its environment. An ecosystem is any place or time where organisms interact with their environment. Ecosystems can be divided into the life in the area and the area in which life exists. Biocoenosis is the life in the area and the biotope is the environment by which life is influenced. One form of interaction in the ecosystem is the food chain, a system moving energy and matter through organisms and the environment. A microecosystem is a small system like a fish with parasites. A mesoecosystem could refer to the coral reef in which the fish lives and a macroecosystem might be the area in which the coral reefs are located and how they interact.

Abiotic factors include geographical, climatological and geological influences. Biotopes are determined by certain abiotic factors. Examples of abiotic factors include, water, air, soil, pH, salinity, temperature, amount of light, and even natural catastrophic events.

A grouping of populations of plant, animal, or microscopic organisms is referred to as a biocenose. When the number of organisms in a species becomes too small, inbreeding reduces the genetic diversity causing weakness in the species and possible extinction. The stability of a biocenose is also connected to biotic ecological interactions by organisms of the same species or organisms of different species.

Interactions between organisms of the same species are cooperation, competition, territorial divisions, and organization in the population. These factors are collectively referred to as intraspecific relations. The interaction of organisms of different species is referred to as interspecific relations and can include symbiotic interactions, competition, parasitism, and infectious disease. When two organisms occupy the same ecological niche, competition for resources can occur. It is important to determine whether the interaction is positive for species, negative for both, or positive for only one species. Another example of an interspecific relation that is negative for one species and positive for the other is predation. Predatory activities form the basis for all food chains. While predation is natural and provides nourishment for the predatory species, in some cases it can upset the balance of the food chain when the prey species is already overexploited.

Biogeochemical cycles are present when minerals and organic materials are used by organisms and sent out as waste. Ecosystems can remain relatively stable when untouched by catastrophic events, detrimental human activities, or other unusual occurrences. Homeostasis, or self-regulation of ecosystems, occurs when supported by natural control mechanisms.

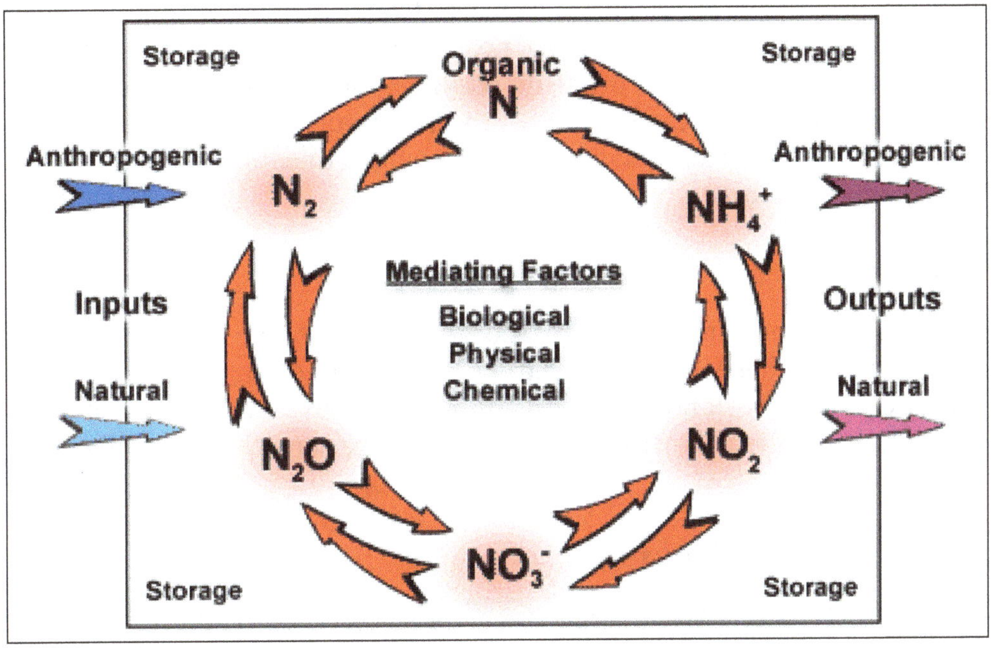

A biome is an ecological area separated from other areas by certain definitive characteristics and relies on the interdependent nature of ecosystems. Water, among other elements, can cycle from one ecosystem to another. Organisms like salmon and freshwater eels often move from one ecosystem to another. The biosphere includes all of the Earth's biomes. An example of a biome is the photic portion of the ocean where sunlight is present and photosynthetic algae can be found. Biomes are divided into ecozones corresponding loosely with the continents and are further divided into ecoregions.

Species are related to each other through the role they play in the food chain as producers, consumers, and decomposers. Producers are photosynthesizing plants, consumers are herbivorous or carnivorous animals, and decomposers are organisms (such as bacteria) that break down organic material into minerals, which are eventually used by producers. There are more producers than consumers. The total amount of living matter in any place is called the biomass. When the biomass of plants increases it is measured as the primary productivity: the biomass produced by consumers and decomposers is measured as the secondary productivity.

Primary and secondary productivity are measurements used by scientists to determine an ecosystem's capacity to support life. An ecological crisis can take place when species or populations evolve in an unsustainable way. Sometimes the quality of the environment is lost after trauma, like a lack of rain or an increase in the temperature of a region. Other times, too much predation can ruin the balance of an ecosystem, as can be seen with overfishing by humans. Sometimes, too many organisms in one place will cause a poor living environment for others.

There are different time scales for ecological crises ranging anywhere from a few months to millions of years. Extinction events can affect many species or just a few individual species. Human activities, like oil spills, can cause local crises as well as global crises like global warming. Even with a local crisis, the loss of a few species can have a disastrous effect on the survival of others in the food chain. A global crisis can result in a loss of nearly all species on Earth. The most common example of such a crisis is the extinction of the dinosaurs. Other examples of well-known ecological crises are the Permian-Triassic extinction event, the Cretaceous-Tertiary Event, Global Warming from the Greenhouse effect, the hole in the ozone layer, desertification and deforestation, and nuclear meltdowns like Chernobyl.

Aquatic Ecology

Aquatic ecology is the study of the plants and animals that live in our rivers and streams and their interactions. These organisms are very sensitive to changes in water quality. Tolerant species (like leeches and carp) are those that thrive in polluted or altered ecosystems while intolerant species (like stoneflies and trout) are sensitive to small changes in water quality.

Phytoplankton and Zooplankton

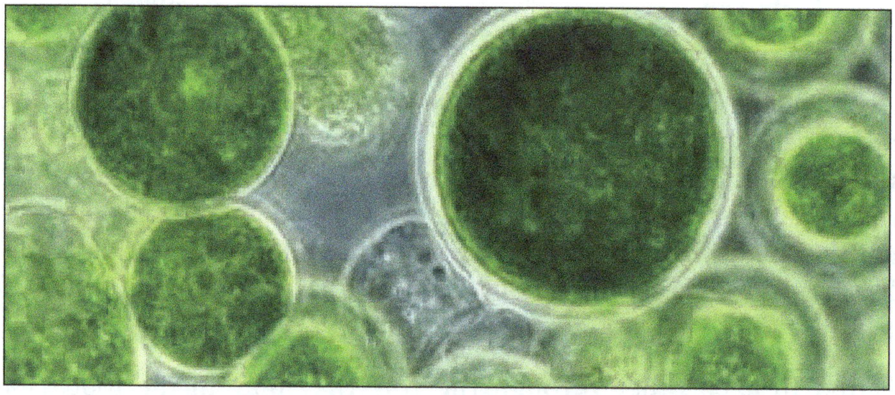

Phytoplankton is tiny plants that use sunlight and nutrients in lakes to grow. Polluted lakes may host 'algae blooms' which can suck oxygen out of the water and may be toxic to humans and animals.

Zooplanktons are small animals that drift and feed on small particles and organisms in the water column. Zooplanktons are the favorite food of many types of fish and most juvenile fish.

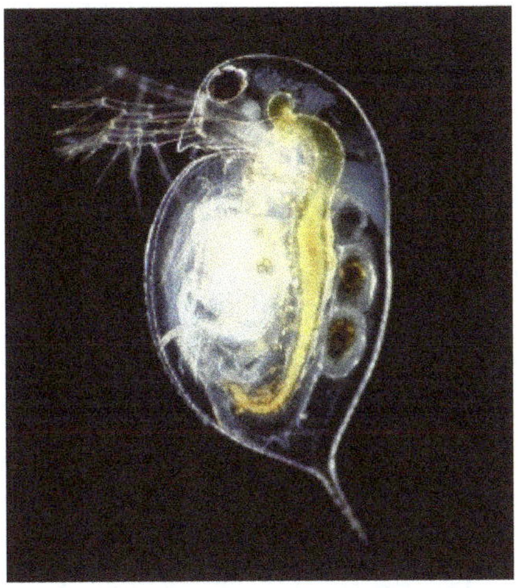

Fish

Fish, besides being a source of food and recreation for people, are important members of aquatic ecosystems and indicators of water quality. Fish occupy different niches in aquatic environments. In streams, fish like the native Bonneville Cutthroat Trout feed on macro invertebrates while in lakes similar sized fish feed on both macro invertebrates and zooplankton. Other types of fish are predators and help regulate the populations of invertebrates and smaller fishes via top-down control.

Not all species of fish can survive in a water body; temperature and water chemistry determine which species of fish can inhabit a water body. For fish, two types of temperatures limit growth and survival. They can survive exposure to warmer temperatures for short periods (hours) but require cooler temperatures over longer periods of exposure. Time of year and age of fish can also affect the temperatures that they require to grow.

Top Down Control

A healthy aquatic ecosystem contains a good balance of producers (like phytoplankton algae and leaves), consumers (like zooplankton and invertebrates) and predators like fishes. The top consumer in an ecosystem excerpts top-down control on the populations of the consumers and producers. In a lake, bluegill suppresses the population of zooplankton via feeding. This then increases the population phytoplankton. This mechanism is known as top down control.

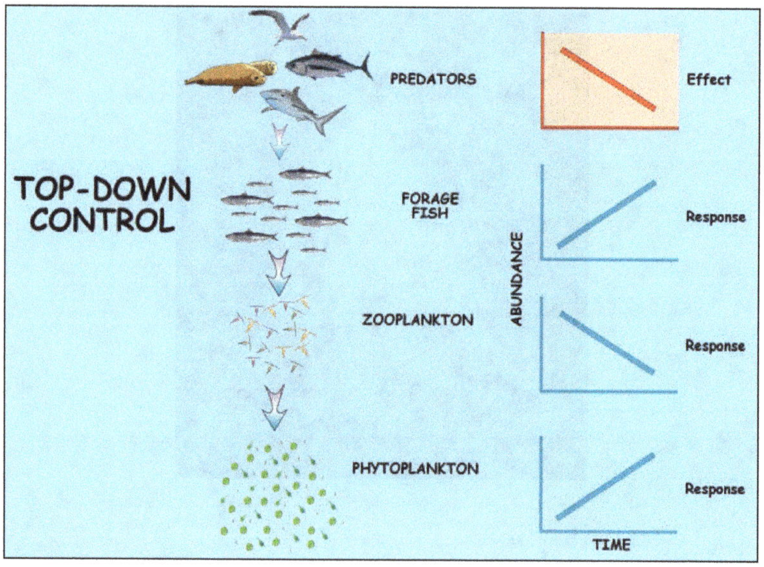

Macroinvertebrates

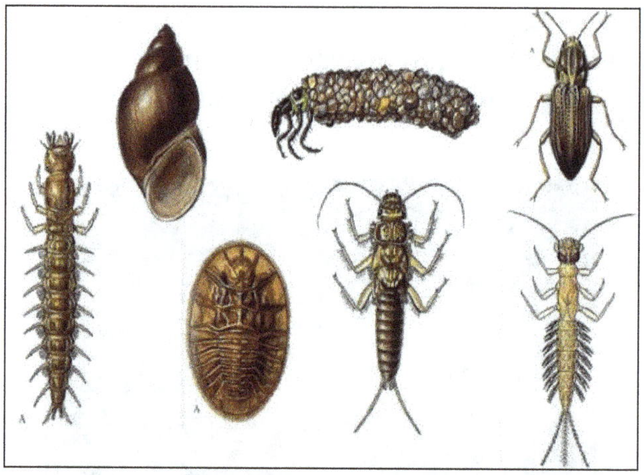

These animals, visible to the human eye and lacking a backbone, include many types of insects as well as other animals such as worms, mollusks, and crustaceans. Macroinvertebrates often feed on plants, algae and things that fall into the stream/lake. In lakes they usually live in the shallow-water area around the edge of the pond amongst aquatic plants.

Macroinvertebrates are often used as indicator species to determine the health of water.Some species are more sensitive to pollution than others. In the pond, dragonflies and mayflies are indicative of a healthy aquatic community.

Types of Aquatic Ecosystem

Different types of aquatic ecosystems are as follows:

Freshwater Aquatic Ecosystem

They cover only a small portion of earth nearly 0.8 per cent. Freshwater involves lakes, ponds, rivers and streams, wetlands, swamp, bog and temporary pools. Freshwater habitats are classified into lotic and lentic habitats. Water bodies such as lakes, ponds, pools, bogs, and other reservoirs are standing water and known as lentic habitats. Whereas lotic habitats represent flowing water bodies such as rivers, streams:

- Lotic Ecosystems: They mainly refer to the rapidly flowing waters that move in a unidirectional way including the rivers and streams. These environments harbor numerous species of insects such as beetles, mayflies, stoneflies and several species of fishes including trout, eel, minnow, etc. Apart from these aquatic species, these ecosystems also include various mammals such as beavers, river dolphins and otters.
- Lentic Ecosystems: They include all standing water habitats. Lakes and ponds are the main examples of Lentic Ecosystem. The word lentic mainly refers to stationary or relatively still water. These ecosystems are home to algae, crabs, shrimps, amphibians such as frogs and salamanders, for both rooted and floating-leaved plants and reptiles including alligators and other water snakes are also found here.
- Wetlands: Wetlands are marshy areas and are sometimes covered in water which has a wide diversity of plants and animals. Swamps, marshes, bogs, black spruce and water lilies are some examples in the plant species found in the wetlands. The animal life of this ecosystem consists of dragonflies and damselflies, birds such as Green Heron and fishes such as Northern Pike.

Marine Aquatic Ecosystem

Marine ecosystem covers the largest surface area of the earth. Two third of earth is covered by water and they constitute of oceans, seas, intertidal zone, reefs, seabed, estuaries, hydrothermal vents and rock pools. Each life form is unique and native to its habitat. This is because they have adaptations according to their habitat. In the case of aquatic animals, they can't survive outside of water. Exceptional cases are still there which shows another example of adaptations (e.g. mudskippers). The marine ecosystem is more concentrated with salts which make it difficult for freshwater organisms to live in. Also, marine animals cannot survive in fresh water. Their body is adapted to live in saltwater; if they are placed in less salty water, their body will swell (osmosis):

- Ocean Ecosystems: Our planet earth is gifted with the five major oceans, namely Pacific, Indian, Arctic, and the Atlantic Ocean. Among all these five oceans, the Pacific and the Atlantic is the largest and deepest ocean. These oceans serve as a home to more than five lakh aquatic species. Few creatures of these ecosystems include shellfish, shark, tube worms, crab small and large ocean fishes, turtles, crustaceans, blue whale, reptiles, marine mammals, seabirds, plankton, corals and other ocean plants.
- Coastal Systems: They are the open systems of land and water which are joined together to form the coastal ecosystems. The coastal ecosystems have a different structure, and diversity. A wide variety of species of aquatic plants and algae are found at the bottom of the coastal ecosystem. The fauna is diverse and it mainly consists of crabs, fish, insects, lobsters snails, shrimp, etc.

Plants and animals in an aquatic ecosystem show a wide variety of adaptations which may involve life cycle, physiological, structural and behavioural adaptations. Majority of aquatic animals are streamlined which helps them to reduce friction and thus save energy. Fins and gills are the locomotors and respiratory organs respectively. Special features in freshwater organisms help them to drain excess water from the body. Aquatic plants have different types of roots which help them to survive in water. Some may have submerged roots; some have emergent roots or maybe floating plants like water hyacinths.

Urban Ecology

The emerging science of urban ecology is a subdiscipline of ecology that examines the interactions between organisms and the human-dominated ecosystems, in which they reside, may provide additional solutions to urban environmental problems. Ecologists first began to perform comprehensive studies of plants, animals, soils, and environmental conditions in cities shortly after World War II, when there were many vacant sites within European cities. They began a tradition of examining open spaces, which supported volunteer plant communities and the animal populations associated with them. Ecological planning emerged as a professional discipline that applied knowledge of the open spaces in urban areas in an effort to enhance biological diversity and amenities originating from green patches in cities.

A different tradition originated in sociology, which applied ecological concepts such as competition and succession to the human communities of cities. In the 1990s, ecological research in urban areas burgeoned. To deal with the complex mosaic of land uses that now make up cities, suburbs, and exurbs, the traditions were combined, along with knowledge taken from other disciplines, to establish a comprehensive ecological approach to the study of urban ecosystems.

Urban ecology has grown increasingly important as a result of the migration of most of the global human population to cities. One of the by-products of this unprecedented phenomenon is that the world's urban areas are expanding into environmentally sensitive locations, where they alter ecosystem structure through pollution and land-use conversion of natural habitats. The knowledge gained from studying biological communities in cities may assist in the development of improved urban design and decision making in dealing with such problems.

Urbanization

Big cities, highly dense population and maximum-imperviousness are local and regional-scale environmental effects of urbanization, which are caused by million-plus, core-oriented, high-rise concentrations. Urban regions are continuing will most probably continue in future to be attraction centers for a number of people. Accordingly, in Europe 75 % of the population live in big cities, 80% will be so by the year 2020. Repercussions of the issue have yet to be grasped within society as a whole. On the other hand, how natural resources are understood, connected and used is profoundly affected by the phenomenon of urbanization. For the great majority of population, various types of urban landscape are 'familiar' environments, and it is expected to be so in future. Landscape is turned into a complex structure by urbanization in terms of forms, materials and activities, which are different characteristics compared to rural landscape.

Human populations living in urban areas cause dramatic effects on the Earth, even though those urbanized parts cover small areas on Earth's surface. The most critical point is that urbanization affects global biodiversity and ecosystems, yet this is not understood adequately. Although there is an increasing interest in urban ecological research, and the understanding of biotic effects of urbanization is better grasped, still, the efforts to bring these issues on the agenda of policy, governance, and planning is lacking.

At the end of the day, urbanization has numerous effects on environment, but this does not mean that all of them are negative. Urban environment is diversified by human effects in various means; this variety of human impact changes ecosystems through urban ecosystems. As a bottleneck, urban areas make environmental changes multiple level issues. Material repercussions of production and consumption modify the usage of land; and affect biodiversity, and hydro systems both locally and regionally; discharge of urban waste impacts on biogeochemical cycles as well as climate at the local and global scales. On the other hand, according to urbanites, the drastic changes in local environment also destroyed global environment. Urban ecology concentrates on this diversity, which is characterized by the interaction between organisms and connections to human doings.

Ecologists did not study urban areas for the first half of the twentieth century, which is worth noting especially considering the concerns of urbanites regarding the future of urban plants and animals. Sukkop claimed that a systematic approach to urban ecological research started after the World War II. Central Europe and the UK can be considered as the pioneers of the oldest urban ecological research tradition. In Berlin, for example, the topic has long been studied. The focus on urban ecology studies has augmented; different arguments on dynamics of urban ecosystems are accrued. It is of vital importance to offer a common action that includes comprehensible definition of ecological conditions inside and outside of cities all around the world.

In 19th and early 20th century periods environmentalists and ecologists neglected the urban landscape, as a result, the 'wilderness' or 'pristine' has been esteemed very highly over the 'human' environments. This perspective approached people as the problem itself; in this sense, removal of people from naturally resourceful areas was the only solution so as to protect the landscape. As a result, cities became the scape goat; they were reasons to destroy nature. Today, this approach is changing; many people are comfortable with the idea that cultural and biological diversity support resilience and sustainability.

One should take into account several urban dynamics in order to understand the relationship between urban sites and biodiversity, and develop a concept satisfying their needs:

- Urbanity stands for city's life quality and character; it refers to the particular pattern of functional, structural, socio-economic and cultural interplay in urban sites. Although many enjoy urbanity, still they prefer to live near nature when they need to make a decision. In this sense, urban revitalization is an attempt to combine nature and city, especially in popular public sites.
- Urbanization is a general term for urban population increase, urban densification and/or expansion and fragmentation of urban sites. It usually leads to increase in a city's ecological footprint, and affects biodiversity and environment as a whole. On the other hand, the nature of footprint depends on the shape and structure of urbanization.
- Urban design is about describing cities' location, physical shape and the construction; therefore, it makes a number of functionalities and life styles possible. In this sense,

through a successful urban design additional urban land-take and fragmentation can be reduced. What is more, urban design can provide green areas, and support biodiversity in cities. Additional green areas, recreating brown fields, promoting 'green' roofs and walls, and keeping the density and compactness in urban areas will maximize ecosystem activities and ecological footprint in big cities. In point of fact, by providing convenient pattern and activity plan, urban area can become more environmentally friendly areas; rather than mere threats to biodiversity.

It is important to highlight the need for an international frame of thought regarding urbanization, which has recently been experienced as a massive, unplanned course of action in landscape change in the world. Urbanization offers a diversity of altered types of land covers in residential, commercial, and industrial areas; they are generally interconnected by roads and railways, on which special green spaces are allocated. This diversity and similar structures are common all around the world; yet; how they affect biodiversity and ecosystem processes has yet to be determined. What is more, international and comparative research attempts are essential in order to develop the understanding of ecological effects of urbanization. Sustainability of future can be maintained only if an ecosystem oriented approach is adopted in terms of urban planning; this approach should include equitable access to ecosystem services and proper planning.

The Necessity of the Ecological Areas in the Urban Landscapes

The English Garden.

The biodiversity in the cities are very much influenced by the presence of green spaces, green roofs and walls along with tree-lined streets. In order to maintain the different ecosystem services including wildlife and human populations, existence of landscapes with sufficient size, diversity and distribution is essential. The green infrastructure in the urban landscape consists of recreation parks and gardens, unmanaged natural open spaces, wetlands and rural lands. According to EEA, the quality of life for most urban dwellers is closely related to the amount of green areas where they live. Since these areas provide people with opportunities for social relations, recreation and experiencing nature which affect them both emotionally and physically. The structure of green spaces and biodiversity are quite different from each other in rural and urban areas as in Europe. Although different counties have different green zone policies, it is accepted that nature affects humans in a very positive way by reducing blood pressure, improving cognitive abilities and increasing happiness. Meanwhile, green areas and biodiversity are also beneficial in filtering particles, purifying water, reducing noise, and buffering climate extremes like heat waves. This means that the green areas in the cities help to reduce the temperatures which are especially important in adopting the future climate changes for the landscapes which are likely to experience high temperatures and heat waves. The English Garden in Munich, Germany is a good sample for this subject.

The presence of natural flora in the cities, the establishment of habitats suitable to the animals adapted to the urban conditions and ecological studies for the protection and development of ecosystems are considered as environmentally ethical studies, as they protect the natural resources. Meanwhile, these areas are places which are increasing the quality of life in the city and allowing the social interaction of the residents.

Ecological studies and designs are significantly important for the sustainability of the mankind and natural environment. The ecological studies carried out for the establishment of the sustainability in the urban areas as well as the for the protection of the resources involve not only the construction-scale efforts performed in the urban areas but also the works conducted in the rural areas. In the construction-scale, the studies comprise efforts such as taking precautions for the extensive use of solar energy, recycling of the domestic waste etc.

Urban Climate

The examination of the climatic conditions for lands differs from that of water. The complex combination of the elements of the lands, such as topography, land form, water and plants, strongly influences the climatic conditions related with solar radiation, wind precipitation, temperature and humidity. For example, the movement of air masses are strongly influenced by land forms and their topography, which are basically considered as obstacles (when they are tall) or routers (when they are in the valley-shaped) for air flow. A variety of air flow dynamics, from simple to complex, could be observed in many circumstances. Not only airflow but also humidity, temperature and absorption of solar energy are strongly affected by the properties of landscapes. The windward and leeward sides of hills affect the humidity, while height of the landscapes strongly influences the temperature. Additionally, higher levels of solar energy are absorbed at the south-facing slopes, giving small micro-variations in the climate and changing the vegetation patterns.

Cities are generally warmer than open lands and forests as walls and roofs of the buildings and asphalt pavements have higher radiation surface than open lands yielding higher amount of solar energy absorbed. Meanwhile, the precipitation falls into the cities flow away to the sewage system through asphalt roads and squares quickly. As a result of this, the solar radiation is more effective on these surfaces than humid ground in the open lands and heat up more yielding higher degree of warming. As the terrestrial radiative heat loss from these city structures is slower, the temperature of the cities is higher. Meanwhile, evaporation of the water in the open lands also results in energy loss. For these reasons, in calm weather and invariable weather conditions, the temperature of a city could be 0.5-1.5 °C higher than that of its surrounding landscape. This difference could reach to 4-5 °C in the night period and even up to 10 °C in the first few hours of the winter nights. Additionally the minimum temperature values could be 0.8 – 1.5 °C higher than what is observed in the rural landscape. As a result of this a city can be termed as heat island. The number of frost and icy days are lower for cities. The warmer climatic conditions of the cities allow a long vegetation period for vegetables and reduce the freezing damages especially observed in the frozen nights. For this reason, the plants that do not grow normally in the climatic conditions of the open land of that region could grow in the city. The heat production in the cities and being warmer compared to their surroundings produce a low pressure region in the atmosphere. This causes a continuous flow of a wind from rural areas to the urban ones. The amounts of the dust and artificial gases in the atmosphere present on the cities are much higher than those present on the open lands and forests. Thus, these artificial materials hinder the sunshine arriving to the city. It was determined that the polluted air reduces the solar radiation by 15-20 %. When the plantation is performed in the cities, the level of solar radiation increases as trees filter the polluted air and dust is partially captured by the long trees.

The effects of the elements present in the urban areas on the flow of the energy. As depicted above, a city can be considered as a hot-spot or heat island compared to its surrounding rural landscape. In fact, the climatic characteristics of a city can be considered for smaller scales, such that the effect of each element in the city, e.g., streets, buildings, parks etc., could be evaluated individually.

Table: The variation of the climatic conditions in the cities.

Element	Compared to Rural Environs
Contaminants	
Condensation nuclei	10 times more
Particulates	50 times more
Gaseous admixtures	5-25 times more
Radiation	
Total on horizontal surface	0-20% less
Ultraviolet, winter	30% less
Ultraviolet, summer	5% less
Sunshine duration	5-15% less
Cloudiness	
Clouds	5-10% more
Fog, winter	100% more
Fog, summer	30% more
Precipitation	
Amounts	5-15% more
Days with less than 5 mm	10% more
Snowfall, inner city	5-10% less
Snowfall, lee of city	10% more
Thunderstorms	10-15% more
Temperature	
Annual mean	0.5-3.0 °C more
Winter minimums (average)	1-2 °C more
Summer maximums	1-3 °C more
Heating degree days	10% less
Relative Humidity	
Annual mean	6% less

Winter	2% less
Summer	8% less
Wind Speed	
Annual mean	20-30% less
Extreme gusts	10-20% less
Calm	5-20% more

The following factors affect the temperature of the urban areas:

- The man-made solid-structures, walls, roofs, roads, paved areas etc., have higher heat conductivities, heat capacities and reflection capabilities than natural soils.
- The surface area of these structures with vertical faces increases the total surface area of the landscape giving higher degree of energy, mass and momentum conversion.
- The heat is continuously produced by equipments, such as machinery, vehicles, heating and cooling systems.
- The amount of evaporation and the energy used for that process decreases in the cities effecting the humidity and heat of the urban areas.
- Exposure to long-wave radiation varies due to the pollutants and dusts are given off to the atmosphere by human activity.

Urban landscapes change the direction and speed of the winds coming from surrounding. For that reason, differences occur between air flow on the cities and on the forests as well as open fields. The buildings in cites which are much taller than the average height increase the number of the calm days in the city and worsen the ventilation. These tall structures are the sources of the calm air, increased temperature and vapor pressure. In fact, this has a negative effect on the living conditions of the city resident. However, tall building do not always reduce the speed of the winds, on the contrary, it is stated that they may improve the circulation of the wind. For this, the large brows (facades) of the buildings should be perpendicular and their narrow brows should be parallel to the direction of the wind. Meanwhile, the distance between buildings should be long enough so that they do not block the motion of the wind. Based on the wind profile and wind tunnel experiments, the following characteristics related to wind flow are apparent:

- Depending on the height profile of the building, the speed of the wind increases at the windward brow of the building.
- The highest value of the wind pressure, which is directly related to the speed of the wind, appears at the 2/3 height of the building.
- However, within the 2/3 height of the windward brow of the building, the speed of the wind was reduced, while the turbulent flow forms due to spinning wind flow.
- The calm air at the windward side of the building is separated form that at the opposite brow while it flows away from top and bottom parts which forms a strong air flow at the lower levels.

The effect of the buildings on the wind flow could be explained by three different examples, which represent simple and complex situations separately. The first case investigates the airflow towards an individual building. Here, when the airflow reaches to the building, an obstacle on route the airflow, its speed should increase according to the continuity of flow laws. A 2D model could clearly show that the speed of the wind on the different parts of the building is different being the highest on the windward brow of the building and across the roof. Meanwhile, air is also deflected from the brow down the face of the building. Additional to those, speeds decrease and streamlines of wind spread out on the leeward side, with some descending toward the ground. In the second case, two buildings of similar heights are located close to each other. Different from the first case, the streamlines of fast wind do not descend to the ground but they were maintained by the roof of the building. The direct consequence of this event is the formation of small pockets of calm air between buildings and limited intermixing of the air in between the buildings with the atmosphere. The properties of air in such pockets could be different than that of surrounding atmosphere. In the last case, the most complex one, buildings and streets are aligned such that the structure is similar to that of canyon forming channels for airflow when the wind is strong. This formation, apparently, increases the speed of the wind and yields turbulence along the walls of the canyon.

The flow of air in an urban area: The figure demonstrates the airflow around different structures, such as tall and short buildings, streets and their different combinations. Different cases are given here: (A) The strong flow is deflected down the building, (B) calm zone develops between buildings and (C) combination of large buildings with streets form canyons yield accelerated airflow.

Other than windflaws, water also significantly affects the urban climate in a way that solar energy is stored by masses of water. As the temperature of water varies much slowly than that of lands, they adjust the temperature of the lands through breezes. Meanwhile, the evaporation of water consumes energy and reduces the temperature of air, which might be called as a natural-air conditioning.

The Urban Heat Island as a Climatic Effect of Urbanization

There are different climatic effects of urbanization: formation of urban heat islands, increasing cloudiness, fog, dust, precipitation and downwind urban heat plume, while decreasing humidity. Apart from these small-scale effects, activities held in the boundaries of urban areas are the main sources of greenhouse gases, mainly CO_2 and fluorocarbons, which may influence the global climate and sea levels in the following decades.

Among different climatic effects of urbanization, formation of heat island is one of the most apparent one. It is known that the temperature of the air in the cities is relatively higher than their surrounding rural areas, defining the urban heat islands. Man-made structures, i.e. buildings, paving, vegetation and other physical elements of the city, form as active thermal elements between the ground and atmosphere. The properties of the elements in the land-covering layer, e.g., height, distribution, determine the thermal behavior of the city and the distribution of the warm air within the urban region.

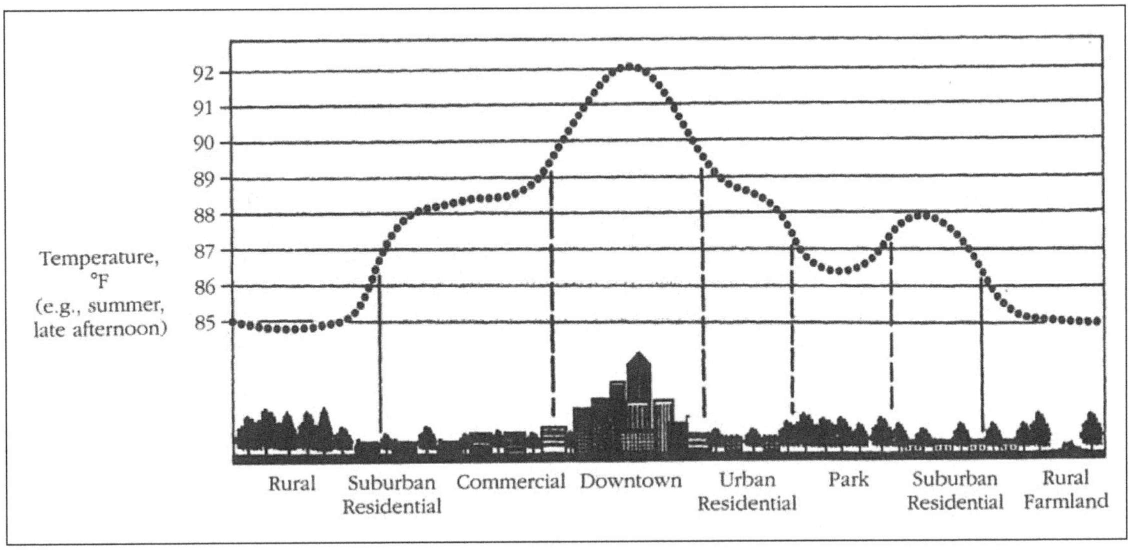

The schematic representation of the temperature profile of the urban heat island and its variation with the type of the elements in the urban area.

There are many factors affecting the formation of the thermal heat islands, primarily, radiation and beat flows, urban boundary layer, surface heating and heat loss. A sharp decrease in the temperature profile occurs on the perimeter of the city, defining the boundaries of the urban heat island by a cliff in the profile. While the temperature may fluctuate in the heat island, which is a factor of the type of the man-made structures, beyond the boundaries of the heat island, rural landscape with much lower temperature values than those observed in the heat island emerges. Such variations in the heat island in a way that cool regions formed by parks and riverside while warmer parts appear as a result of the shopping centers and industrial zones.

According to Leopold, the use of the local-land has four distinguishable effects on the hydrology, which are relevant to each other: changes in peak-flow characteristics; changes in total runoff; changes in water quality; and changes in hydrologic amenities. It is known that stream flows, which are rainstorms characterized by heat islands, expands and becomes stronger with the growth of the cities, while stronger winds are necessary to overcome them.

In fact, a lot of urban planners accept climate as a fundamental factor of urban planning and design; on the other hand, only few of them are able to combine variables like wind, heat, and solar radiation with decision making processes. The level of scientific understanding of microclimate is one of the most significant factors responsible for building the environment. Climate is an important element for pollution control boards and regional transportation policies; whereas, for urban planner, landscape architect, and architect small-scale climatic variations are important. One can refer to the solar radiation around high buildings as an example of variation; other parameters like temperature, wind, fog, and pollution can also be seen as other variations in urban landscapes.

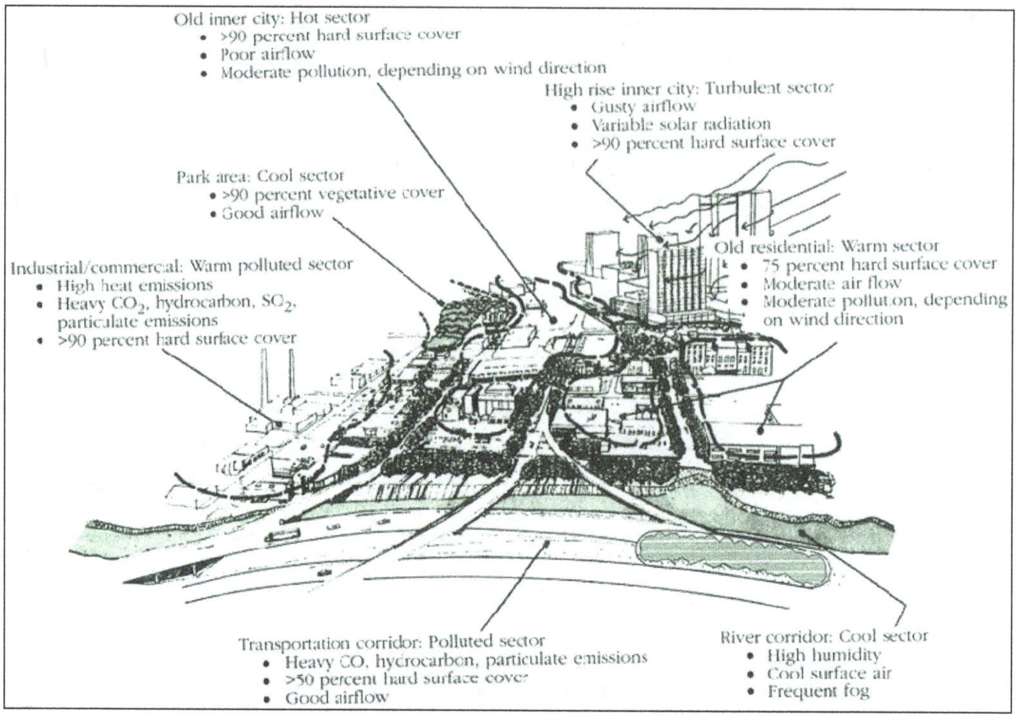

Climatic conditions near the ground in different sectors of a city. Conditions vary with surface cover, solar radiation, airflow, and air pollution, among other things.

Urban Hydrology

In urban design, groundwater relations and surface flow characteristics of water collection basins are of vital importance for the city's water requirement. Water economy affects heat economy as well. In cities, precipitation falling on building roofs and asphalt roads as well as city centers, are all lost because they go to sewer system through surface fall. As a result, inside the cities get dry and warmer very soon. Because precipitation waters are carried outside of the city, the level of humidity in cities is lesser than forest design. However, fog comes down frequently due to high quantity of solid structures.

Within the framework of urban hydrologic processes, trees as well as soils can act as an important role in terms of intercepting and retaining or slowing the flow of precipitation reaching the ground. It is possible that these can decrease the frequency and volume of storm water runoff, flooding damage, storm water treatment costs, and other problems in terms of water quality. According to the runoff estimates, which were conducted for an intensive storm in Dayton/OH, the tree canopy (22%) decreased runoff by 7%, and a decent increase in canopy (around 29%) can decrease runoff by almost 12%. In addition, it was revealed that tree cover on pervious surfaces decreased runoff, total of 40 %; its effect on runoff was lesser on impervious surfaces.

In Europe, according to the findings obtained from research, low level humidity and more frequent fogs cause torrential rainfall repetition to increase. For this reason, it was determined that the number of days of torrential rainfall in the city is between 13 and 63% more than the open area and forest designs. Therefore, it was revealed that the city's annual rainfall is around 5 – 31% higher than open area and forest designs, depending on the land surface shape, height, climate characteristics, the size of the structures. However, torrential rainfall and impervious surfaces inside the city cause this precipitation water to be lost in a short time; rain water does not contribute to the water economy efficiently.

Increased impervious cover is among the most significant modifications affecting streams in urban sites; it transforms hydrology and funnels accumulated pollutants from buildings, roadways, and parking lots into streams. In United States point source pollution was reduced by regulations; yet it is still an important problem in several developing countries. It is crucial to remember that industrial discharges and sewage contaminate both rivers and lakes. In new design cities storm water infrastructure is separate from wastewater discharges; this is not the case in older European and American cities, where two streams are mixed and causing to acute pollution in recipient systems. The storms and low flow-discharge of cities lead to local, even regional pollution downstream, which are particularly caused by pesticides and persistent organic pollutants.

Urban hydrology has been altered to develop a way to manage urban runoff against flood, and to protect health and the environment. In recent years there have been important developments regarding the measurement and prediction of urban rainfall by using technologic devices such as radar and microwave networks. The predictability of urban hydrology has been improved also in order to provide models convenient with small temporal and spatial scales typical to urban and pheri urban applications. Urban storm water management has started to take into account the needs of environment and human beings. There is a strict trend towards restoring pre-development flow-regimes and water quality; also, it is believed that restoring more natural water balance would not only be an advantage to environment, but support the livability of the urban landscape. Although it was regarded as a trouble, today storm water is increasingly seen as a resource.

Urban runoff management has evolved with the understanding of its environmental impacts. Referred approach developed many terms such as, Sustainable Urban Drainage Systems (SUDS), Water Sensitive Urban Design (WSUD) and Low Impact Development (LID). These approaches have similar objectives:

- To manage urban water cycle in a sustainable fashion (by taking into account surface and ground water, flooding and impacts on waterways erosion).
- To maintain or return the flow regime as much as possible back to its natural level.
- To protect and restore, if possible, water quality (for both surface and ground water).
- To protect and restore, if possible, the health of receiving water.
- To conserve water resources as a resource rather than a nuisance (like, storm water).
- To enhance urban landscape and amenity by incorporating storm water management measures, which offer multiple benefits, into the landscape.

Urban Soil

Analyzing and determining the soil features are important in urban design for two reasons. First, it is important in terms of geologic basis for structure projects like edifice, tunnel, subway, etc. Second, it is mandatory to know the soil characteristic in growing plant. Physical features of soil, which can be listed as depth, mechanical structure, general characteristics, level of ground water, and geological basis; are not only important for tunnels and edifices, but also for all urban structures (road system, sewage, cesspool, etc.). Soil is the natural environment for plants. Hence, determining soil features is of particular importance for the planned green areas in the city. For this reason, in establishing cities, favorable soils for structures should be found through soil analysis and marked on maps, and the relations between soil characteristic and how to benefit from the soil should be analyzed.

The soil in the areas that were left to make greenery near the building has lost its genetic characteristic and mixed with construction leftovers, cement, and concrete. Therefore, to those spots, on which green areas will be built, soil from other places are carried on trucks; they do not include concrete, but organic materials; and they are generally spread in a thin layer. For this reason, natural soil features, which are main factors of growing plant in cities, have rarely survived. As for the plants that will be grown in cities, shadows of buildings and windbreak, negative effects of salts sprayed onto the city roads in winter, and flue gas should be prior factors to consider. In this sense, plant species that will be used in urban design should have a broad ecologic tolerance; in other words, they should be durable against negative anthropogenic effects.

Urban Biodiversity

Biodiversity provides an ecosystem system, on which the standard of living on European citizens depends, urban setting proves this successfully. Urbanites benefit from recreational, social and inspirational services of the nature, both inside and outside of cities. Yet, ecosystems are also crucial in terms of basic living conditions in cities. Thanks to its positive climatic effects, urban greenery will play an important role in adapting strategies to deal with climate change, which is expected to intensify or alter the specific urban climatic conditions. Urban greenery makes cities attraction centers; it prevents urban sprawl and saves space for biodiversity. In addition, as in-city services increase, the city's footprint is decreased; as a result, potential negative effects on biodiversity and environment are destroyed.

In addition to ecologic services, green areas –in which biologic diversity is supported are also gaining favor in terms of public health, maintaining social cohesion, economic benefit (like the increase in real estate market thanks to the green areas), and low maintenance cost (less irrigation and fertilization by creating ecologically sustainable urban designs).

Local and regional authorities have the legal power to designate conservation areas; to advocate EU's Natura 2000 networks; and, to bring biodiversity concerns urban and spatial planning agendas. Cities share the responsibility to select Natura 2000 sites. The scale of public commitment can be traced in Local Agenda 21 processes, which aim to establish sustainable societies identifying biodiversity as a precondition to build resilient cities. European Union supports advanced commitment and increased awareness by honoring the most sustainable cities, the European Green Capital Award, and establishing a legal framework to protect biodiversity through, Natura 2000 network under the EU Habitats and Birds Directives, Air Quality Directives, the Water Framework Directive and the development of a Soil Directive.

Cities are crucial in terms of sheltering some rare and endangered species and habitats, which are considered to be important at the European level as well. There are a total of 97 Natura 200 sites are located in 32 major European cities. Of these cities, sixteen are capitals (for example, London, Paris, Prague, Rome and Tallinn). In addition, one or more harbors in more than half of the EU's capitals are Natura 2000 sites. Although Berlin has 15 Natura 2000 sites, most of the others have one or two. A new City Biodiversity Index (CBI) has been developed by favor of Convention on Biological Diversity (CBD) and contributions of Governments for Sustainability (ICLEI).

In cities, the value of biodiversity is closely related to societies' cultural and social preferences. Biodiversity not only offers quite an amount of ecosystem services to urbanites, it also modifies negative perceptions created by cities.

It is possible to perceive urban diversity within the framework of response diversity, which is the case when species and ecosystems respond differently to external interventions, even though they function collectively. For instance, urban trees intercept large amounts of precipitation, and prevent flooding; therefore, tree diversity contributes to precipitation interception's response diversity function. As a result, strong biodiversity guarantees lower risk for the entire function, despite local extinction. As the cities continue to face the effects of climate change, response diversity will be an important capacity building factor in terms of resistance.

From the perspective of urban planning, protection of biodiversity in cities is a high-priority for both stakeholders and decision-makers. Still, when it comes to put the issue at the top priority list during the planning process, it is left behind of anthropocentric objectives such as economic development, transportation, land use and recreation.

Extensive use of native species is essential for the sake of urban biodiversity. By doing so it could be possible to carry out more successful landscape projects with minimum maintenance work and cost. Furthermore, a breeding and feeding environment can be created through increasing biodiversity and by using native species. Regarding sustainable cities, floristic diversity is an important topic.

Biotic, physical and social factors along with processes are related to biodiversity in urban ecosystems. On the other hand, ecology itself is insufficient to delineate urban ecosystems. As a result, it is advised to establish an interdicplinary approach, including natural and social sciences that will discuss ecology-related inputs in terms of urban planning.

Urban Wildlife

Wildlife in cities is generally undervalued. In recent years urban wetlands, abandoned industrial sites, roadside verges, vacant lots and derelict lands, ruins, allotment gardens and cemeteries - together with arboreta, residential gardens and villas, botanic gardens and individual balconies have begun to be seen as potential conservation areas for urban biodiversity. Dense population and infrastructure are two factors putting pressure on biodiversity in cities. What is necessary is to establish a balance between the urban green and tailor made green areas. At this point, 'double inner city development' can be used; this means to combine the existing 'made' areas with conservation, supporting the presence, character and availability of green spots and vacant spaces; and, strengthening the green infrastructure like street-trees, green walls and roofs. These will facilitate the way to access green areas in and out the city. Wild fauna inside the cities can be a positive demonstration of the richness of the 'green'; on the other hand, this can be a challenge in some cases (e.g. foxes or seagulls).

Rich diversities in urban landscapes can come to the fore as original communities. Urban green is an important part of urban landscape, which offers the opportunity to contact with wildlife in addition to environmental and socio-ecological benefits, regarding the quality of human life. Urban green areas are ecologically complex structures; their values can be defined in terms of goods and services within society. What is essential is to pick bio-indicators that can be accessed and relied with regards to create a proper/balanced urban ecosystem, through integrating wildlife and biological parameters to human well-being.

Ecology by Scale

Microecology

Microecology or Microbial ecology is the study of the interactions of microorganisms with their environment, each other, and plant and animal species. It includes the study of symbioses, biogeochemical cycles and the interaction of microbes with anthropogenic effects such as pollution and climate change. Most types of microbes remain unknown. It is estimated that we know less than 1% of the microbial species on Earth. Yet microbes surround us everywhere -- air, water, soil. An average gram of soil contains one billion (1,000,000,000) microbes representing probably several thousand species:

- Archaea: Single cell microorganisms.
- Bacteria: Can be found in soil, water and the Earth crust. They also live in symbiotic and parasitic relationships with plants and animals.
- Fungi: Microorganisms such as yeasts and molds, as well as the more familiar mushrooms.
- Protists: Primitive algae, amoebas, slime molds and protozoa.

Viruses

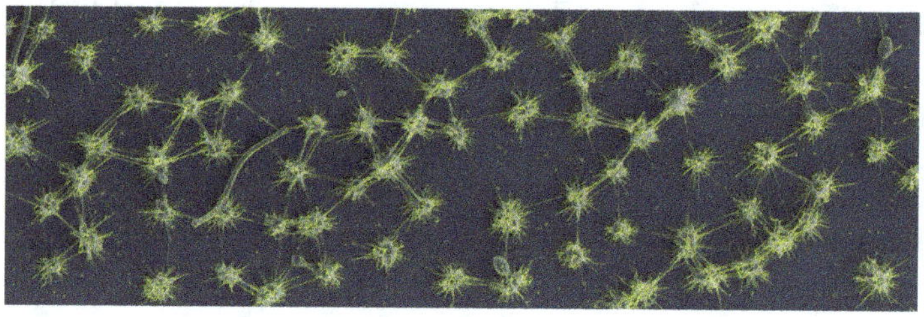

Viruses are small infectious agents that replicate only inside the living cells of other organisms. Technically they are not living microbes but they affect microorganisms, including bacteria and archaea so this is why it is included in the study of Microbial Ecology.

Importance

The study can help us improve our lives via the use of microbes in environmental restoration, food production, bio-engineering of useful products such as antibiotics, food supplements, and chemicals. It helps measure the effects of climate change and land usage.

Biodiversity

An estimated 1,000,000 bacterial species exist on this planet, according to the Global Biodiversity Assessment, yet fewer than 4500 have been described. The greatest genetic diversity of life comes from within the world of microorganisms, yet the least is known about them.

Diverse Habitats

Microbes inhabit the widest range of habitats from sub-freezing temperatures, to water hotter than boiling, from the rocks beneath our feet, to the atmosphere miles overhead, to the stuff between our toes, to the tops of mountains and to the deepest ocean trenches.

Microbe Hunting

Hunting for new microbes is not as easy as taking a jeep trip in the outback with a pair of binoculars. By definition, microbes are invisible without the aid of a microscope, so the challenge to find new ones is great. The difficulty is compounded when one does look under the microscope to see two apparently similar bacteria which later prove to be not at all similar. For example: two bacteria may have the same rod shape, but one thrives in the presence of oxygen whereas the other one is killed by oxygen. Because microbes are so difficult to observe, they are the last organisms to be catalogued with less than one percent yet described.

Macroecology

The idea of macroecology as a distinct field of research has been around for more than two decades and was conceived as a response to the realization that small-scale local processes alone were not able to fully explain the abundance and distribution of species. This led to a broader perspective that searched for generalized patterns at large spatial and temporal scales, characterized by the search for statistical relationships to explain the distribution of biodiversity from a historical and geographical perspective. Ten years ago, a symposium of the British Ecological Society (BES) was convened with the aim of reconciling divergent perspectives on large-scale ecological patterns. This 'Causes and Consequences' symposium set the tone for a decade of research in macroecology.

From Pattern to Process

The strongest theme that percolated all of the talks was the increased emphasis on the processes that drive biodiversity patterns). This theme was introduced by Owens, who outlined a shift from describing patterns to a search for mechanistic understanding. This theme was further developed by Sean Connolly, who identified a mismatch between the biological reasoning that underpins hypotheses about the drivers of macroecological patterns and the statistical models that are actually fitted to data. Connolly illustrated how this has hindered progress in our understanding of large-scale species-richness gradients, and demonstrated how models based on biological processes can be used to derive testable hypotheses. Although macroecology is relatively advanced in its use of statistical methods, the theoretical basis of the predictions involved is sometimes poorly developed. Connolly argued that the explicit formulation of theoretical models, and the robust derivation of statistical expectations from those models, is one of macroecology's most significant challenges.

Katrin Böhning-Gaese provided a clear demonstration of how incorporating local processes can influence large-scale patterns of species distributions. For example, projections of the impact of climate change on bird species richness yielded very different results when biotic interactions with tree species were taken into account. Similarly, Trevor Price emphasized that both biotic and abiotic factors can explain large-scale diversity gradients. He showed how niche conservatism is not enough to explain diversity gradients of Himalayan birds, unless competitive interactions are incorporated. Kate Jones and David Redding showed how the spread of a zoonotic disease (Lassa fever) can only be understood with reference to the distribution of the host (a rat).

Moreover, Nicholas Dulvy described how the thermal tolerance of individual organisms underpins the distribution of poikilothermic animals in the oceans, and their responses to recent climate change, but that this was not the case on land. Dulvy speculated that gross differences between marine and terrestrial environments can be attributed to the importance of behavioural thermoregulation and interspecific competition on land, contrasting with the dominance of size-based competition in marine systems.

The increasing focus on mechanistic understanding in macroecology is not confined to this meeting and many of the recent attempts to build unified theories in ecology have been process-based. A key challenge now is to derive general and testable predictions via robust theoretical modelling, underpinned by biologically reasonable assumptions. Recent progress in this area has been substantial, although many current theories may not be testable even for data-rich taxa, such as mammals. Thus, further research to bridge the gap between theory, predictions and data is a priority for the development of macroecology into the future.

Breaking Down Barriers

Traditionally, macroecology focused on processes operating at large scales (e.g. climatic and phylogenetic), largely ignoring the potential for small-scale processes to generate a coherent signal in macroecological patterns. One reason is the deficit of fine-grained (e.g. population-level) datasets that are replicated over large spatial extent: national monitoring schemes have great potential in this regard. A growing body of evidence, both theoretical and empirical, suggests such signals can be detected. Conversely, Böhning-Gaese showed large-scale abiotic gradients can influence community assembly. One striking example is that the degree of specialization, identified using interaction networks among pollinator and frugivore species, is greater in temperate than in tropical communities, contrary to expectation. Böhning-Gaese argued that advances in understanding how ecological patterns are generated at multiple spatial scales, and how they are interrelated, are important steps towards a multi-scale synthesis across ecology.

An additional barrier to progress within ecology in general is the lack of synthesis across taxonomic groups and biomes. Historically, macroecology was no exception, being predominantly focused on terrestrial vertebrates, although marine macroecology was well-represented at this meeting. A feature of the presentations by marine ecologists was that the concepts and analyses they use are not exclusive to the marine environment. Connolly's process-based models of species richness are wholly transferrable to terrestrial cases. Contrasts between realms can discriminate among hypotheses. For instance, equator-ward range limits on land were previously explained as an artefact of under-sampling in the tropics, but the contrast with changing marine range limits in the tropics, where scientific capacity is also low, suggested that stagnant terrestrial ranges are real. More generally, inter-realm comparative analyses provide many novel opportunities to test mechanistic macroecological hypotheses.

New Applications

The meeting demonstrated well how macroecology has influenced diverse research agendas, further reinforcing its application to public policy on biodiversity. Owens argued that the influence of macroecology has been unusually broad and deep at the interface of science and policy, especially around land-use, climate change and biodiversity loss. Thus, a significant opportunity exists for macroecology to remain influential and adapt to changing priorities of stakeholders and funding bodies. Two talks focused specifically on the extent to which macroecological ideas are gaining traction in mapping ecosystem services (MES) and epidemiology.

MES, and the potential trade-offs among them, is ripe for the application of macroecological approaches. Like macroecology, MES examines correlations in space over large scales, for example, calculating the degree of spatial overlap of multiple services. Felix Eigenbrod argued MES should adopt macroecological tools to identify the mechanisms underpinning the distributions of ecosystem services. A further challenge for MES lies in the necessity to consider linkages between the distribution of biophysical stocks and their potential beneficiaries, which is somewhat analogous to modelling overlapping geographical ranges of interacting species. For example, Böhning-Gaese incorporated species richness of fig trees (the stock) into predictive models for frugivorous birds (the beneficiaries). Therefore, the incorporation of co-occurrence and subsequent interactions within both research agendas may be an area that would benefit from collaboration.

References

- Ecology, biology: byjus.com, Retrieved 15, April 2020
- Community-ecology, environmental-studies, ecology-and-environmentalism, earth-and-environment: encyclopedia.com, Retrieved 05, May 2020
- Ecosystem-ecology: courses.lumenlearning.com, Retrieved 03, June 2020
- Systems-ecology: encyclopedia2.thefreedictionary.com, Retrieved 07, April 2020
- What-is-system-ecology: greennet-project.eu, Retrieved 12, January 2020
- Landscape-Ecology: biologyreference.com, Retrieved 10, July 2020
- Animal-ecology, encyclopedias-almanacs-transcripts-and-maps, Animal-ecology-concerns-the-relationships-communities-energy-flow, social-sciences: encyclopedia.com, Retrieved 13, June 2020
- Introductory-chapter-plant-ecology, plant-ecology-traditional-approaches-to-recent-trends: intechopen.com, Retrieved 05, August 2020
- Microbial-ecology, boundless-microbiology: shreeichhamanifilms.com, Retrieved 19, February 2020
- Forest-ecosystem-types-characteristic-features-and-structure, ecosystem, environment: yourarticlelibrary.com, Retrieved 24, March 2020

Chapter 2

Ecosystem and its Components

An ecosystem can be defined as a community or a group of living organisms within a geographical area that live and interact with each other and the environment. An ecosystem consists of both living and non-living components. Ecosystems, their different types and components are discussed in this carefully written chapter to provide a better understanding of the subject.

An ecosystem is an area that includes all the biotic (living organisms) parts such as plants and animals and the abiotic (non-living) such as soil, rocks and climate sharing an environment.

How does Energy Flow through an Ecosystem?

Organisms within an ecosystem can be classed as producers, consumers or decomposers. Energy flows through these organisms within the ecosystem. Producers, such as trees, produce their own food and begin this cycle. Using energy from the sun they produce food. They do this by photosynthesis. Most producers are plants, but there are some small organisms that produce food through photosynthesis as well.

The producers are eaten by primary consumers that cannot produce their own food, such as a giraffe. Primary consumers are herbivores which mean they only eat plants. Secondary consumers are carnivores such as lions. In a simple food chain, secondary consumers eat primary consumers.

Decomposers and break down dead plants and animals. They also break down the waste of other organisms. Examples of decomposers include bacteria and fungi. Decomposers get their energy from breaking down dead material e.g. dead producers, dead consumers or fallen leaves. When dead material is decomposed nutrients are released into the soil. These nutrients are then taken up from the soil by plants. Decomposers are very important for any ecosystem. If they weren't in the ecosystem the plants would not get essential nutrients and dead matter and waste would gather.

A food chain.

A food chain shows the relationships between these feeding groups. They show the flow of energy from producer to tertiary consumer. The image above illustrates this.

A food web shows lots of food chains and how they overlap.

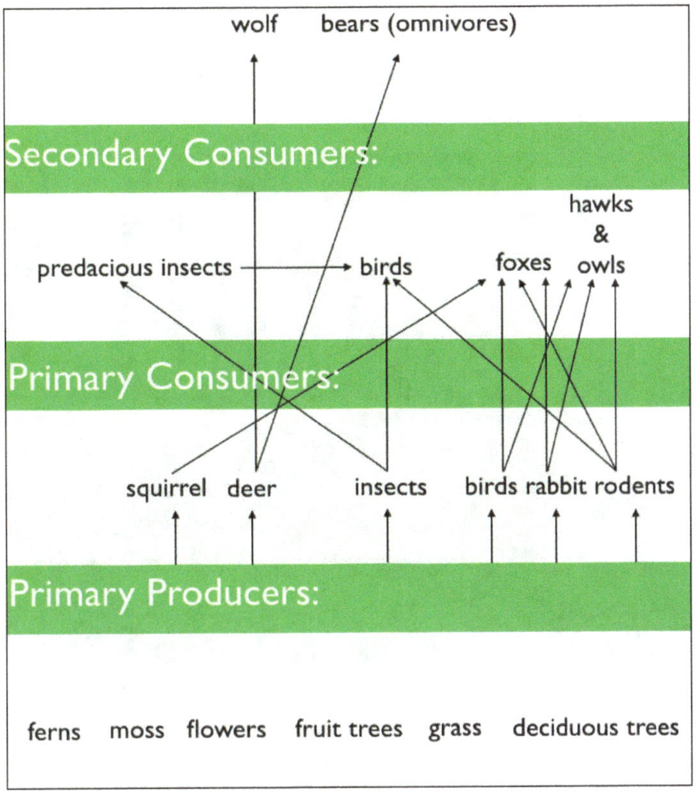

A food web.

The transfer of nutrients through the ecosystem is known as the nutrient cycle. The nutrient cycle shows pathways through which nutrients are constantly recycled. All plants and animals depend on nutrients in food for their survival.

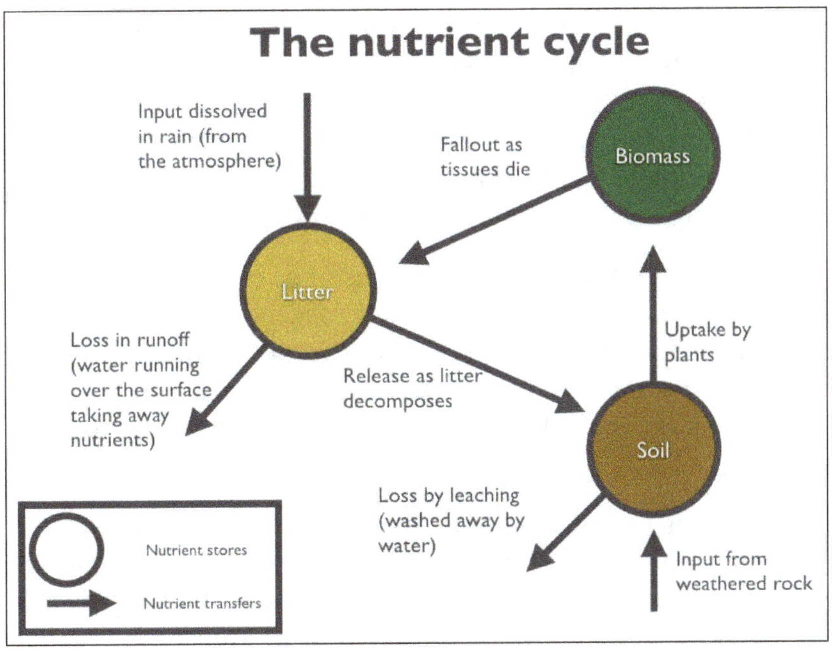

The Nutrient Cycle.

A hedgerow is an example of a small-scale ecosystem. A hedgerow ecosystem includes the plants that make up the hedgerow, the organisms that live in it and feed on it, the soil in the area and the rainfall and sunshine it receives. The producers include hawthorn bushes and blackberry bushes. The consumers include thrushes, ladybirds, spiders, greenfly, sparrows and sparrow hawks. The largest ecosystems are called biomes.

Types of Ecosystem

Ecosystem types abound; however, the most basic categorization involves three habitats: terrestrial, marine, and aquatic. Naturally, these groups can be split into thousands of smaller systems, each one offering a different mix of climate, habitat and life forms.

Terrestrial Ecosystems – Global Land-based Habitats

At approximately 57 268 900 square miles, the terrestrial ecosystem covers just 29% of the globe. As these habitats are varied, terrestrial ecosystems are further broken down into six types. The deciduous forest ecosystem is found in temperate regions and experiences temperature and precipitation fluctuations according to four seasons. Current conservation goals include reintroducing apex predators after the culling practices of previous centuries, and providing an environment full with mature trees to make up for unregulated deforestation.

Desert ecosystems can be hot and dry, semi-arid, coastal or cold. The feature that links these is a lack of water and the absence of a soil layer in which larger vegetation such as shrubs and trees can thrive. While indigenous life has adapted to the absence of water, a desert is still unable to support the populations of a wetter habitat. Substantially sized herbivores are unable to survive in a desert environment in large numbers, and this in turn limits the numbers of larger omnivores and carnivores.

Grasslands are also known as prairies, pampas, savanna or steppe. They can be tropical or temperate, and are a link between desert and forest. They rarely receive enough rain to support trees, but possess enough soil nutrients to feed large, grassy expanses. This provides considerable energy for primary consumers. With a large population of producers, a grassland ecosystem can similarly support large herds of herbivores, which in turn feed consumers higher up in the food chain hierarchy.

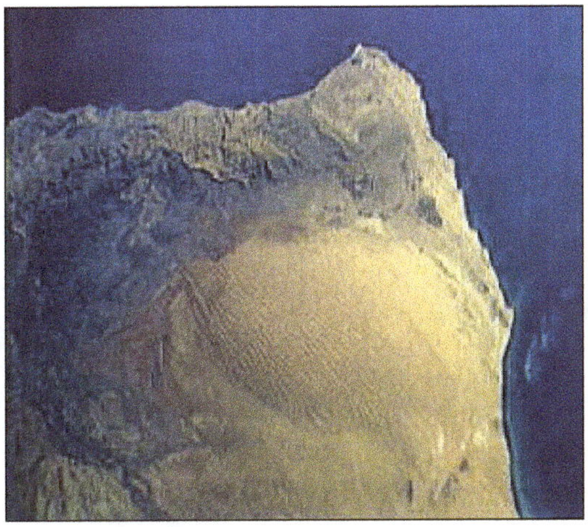

Satellite image of a desert in Oman.

The taiga is a region of subarctic forest south of the Arctic Circle. It has layers of permafrost or rock under shallow soil, which make the soil marshy. The taiga supports huge numbers of conifers – slow growing, cold-resistant trees. Other plant life is small and includes lichen, marshland plants and small shrubs. The map below shows how this ecosystem is distributed across the globe.

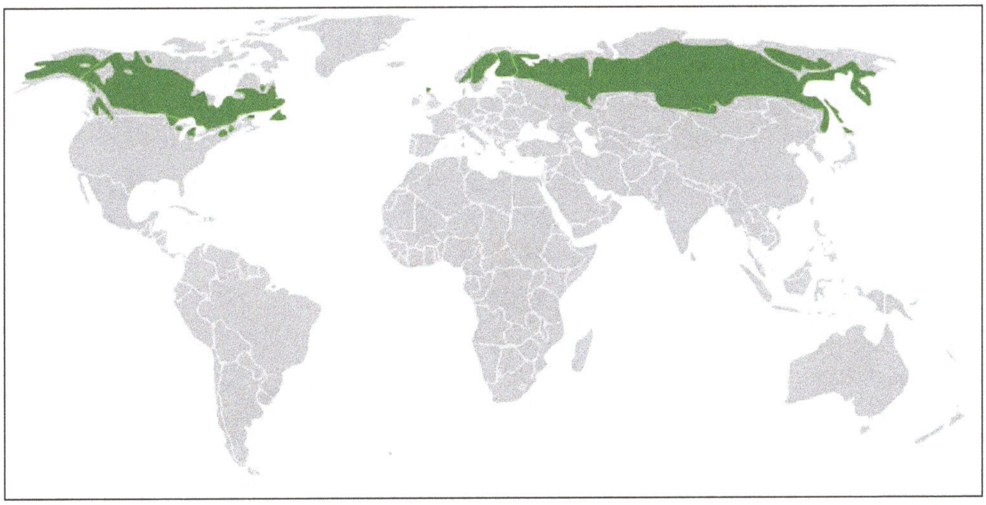

Map of global Taiga ecosystems.

Tropical rainforests are probably the most quoted ecosystems in the field of environmental conservation. Located around the Equator, constant rainfall and warmth together with a lack of seasons provide a stable climate, yet cloud and the tree canopy make the rainforest floor a dark place. Soil is leached of nutrients through constant precipitation. Plant life has adapted and is abundant, making the tropical rainforest ecosystem the terrestrial ecosystem with the most biodiversity.

Tundra, the last of the six terrestrial biomes, is the treeless environment of the Arctic Circle. Climate change is rapidly changing this ecosystem, as warmer weather brings non-indigenous predators in, where they compete for limited prey. Certain shrubs are taking root as the Artic permafrost layer melts. These compete with lichen – the primary food source of caribou.

The Siberian Tundra.

Marine Ecosystems – Biodiversity in Oceans, Seas and Glaciers

Marine and aquatic ecosystems cover 139 668 500 square miles; 97% of this is salt water, making marine ecosystems the largest biome category. Large marine ecosystems (LMEs) are particularly difficult to observe and control, as different salt water habitats have complex chemical compositions that vary from coast to coast and from shallow to deep. These compositions are forever shifting due to tides and currents. Pollutants and organisms travel on courses which, although predictable, are continuously in motion. The sheer volume of water a large marine ecosystem covers is immense. The map below shows the population trends of native and invasive species of jellyfish and their populations. This study looks at the trends of jellyfish behavior in an LME. To predict this trend on a global scale is possible, but the potential and actual variables are countless.

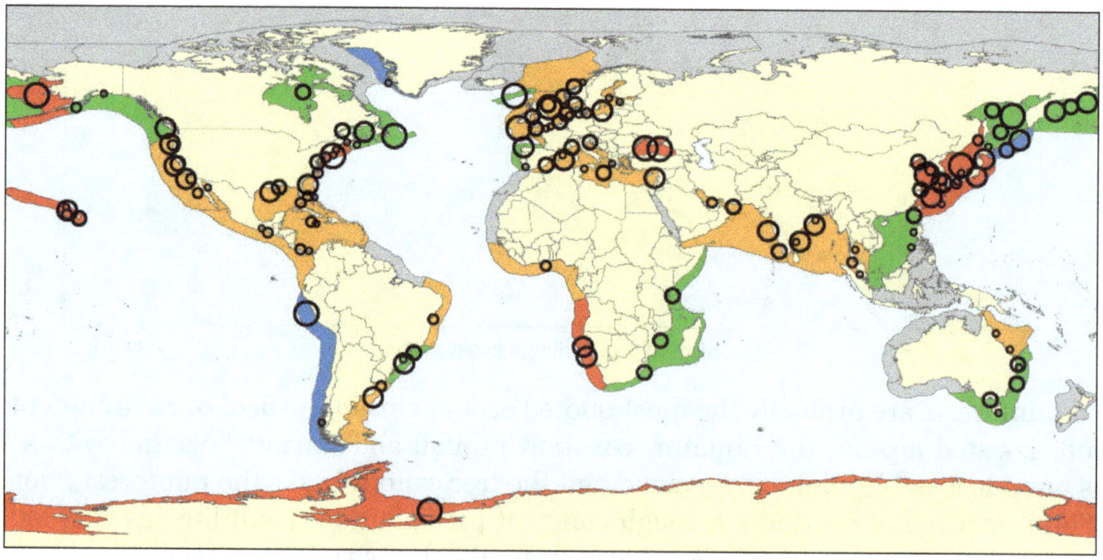

Global jellyfish populations and interactions.

Marine or ocean ecosystems are grouped into open marine, ocean floor, coral reef, estuary, saltwater wetland estuary, and mangrove systems. These cover marine environments from the surfaces and floors of the deepest oceans to partially terrestrial, tidal swamps.

Aquatic Ecosystems – Freshwater Locations: Lakes, Pools and Rivers

Freshwater ecosystems cover approximately 3% of the planet's surface. Aquatic ecosystems also include estuaries before freshwater meets salt, wetlands, ponds (natural or artificial), lakes, and rivers. As fresh water is essential for all life, aquatic biomes are extremely important. Yet they are very small in comparison to other habitats, and have been used as dumping grounds for centuries. National Geographic reports that freshwater species are four to six times more at risk of becoming extinct than terrestrial or marine species. Freshwater biomes and coastal marine systems are also at high risk of eutrophication, a natural process which requires centuries to develop. Eutrophication is caused by increased levels of sediments which in turn increase levels of nutrients and encourage excessive plant growth. When vegetation dies off, after depleting the additional nutrients or becoming victims of their own success, their decomposition leads to dead zones, or hypoxic zones.

Microbiome – Supporting the Biodiversity of Living Organisms

Any anatomical system contains microbiota – mutualistic, commensalistic, pathogenic or parasitic bacteria, fungi, archaea and viruses. A relationship between gut biodiversity and the health of other anatomical and physiological systems, such as mood, hormonal production and resistance. This may be likened to the effect of reducing biodiversity in aquatic ecosystems, and the effect this will have on populations at distant locations. For example, research is looking into the prevalence of chronic inflammatory diseases as microbial diversity in the intestine decreases. The microbiome may not be an ecological ecosystem, but it is a complete ecosystem of living and non-living components in a habitat where interactions take place, and which has its own climate.

Abiotic Components

Abiotic components or abiotic factors are nonliving factors that impact an ecosystem. These factors are part of the ecosystem and influence the associated living things but they are not living. The term abiotic is a mix of two words, these are a- which means without and bio which means life.

What are the Types of Abiotic Components?

There are five different types of abiotic components. These are:

- Temperature: A rise in temperature can change the development in an animal; can cause changes in metabolic activity and much more. All organisms can tolerate a certain range of temperature and how extreme temperatures lead to stressful conditions.
- Water: Water covers more than 70% of the earth's surface in one form or the other. Compared to that, living organisms require a small amount of water to live. Water is critical to survival.
- Atmosphere: The atmosphere has important components like oxygen and carbon dioxide, which animals and plants breathe to live and combine them to produce carbohydrate, other organic materials, part of DNS and proteins.

- Chemical Elements: Chemical elements play a major role within the environment to influence the type of organisms which can grow or thrive in an area. The chemical composition, including pH level, has a huge impact on the plants of an area. For example, plants like azaleas thrive in acidic soils. Some elements, like zinc and copper, are important micronutrients for the development of many organisms.
- Wind: The wind direction and speed in an area affects its temperature and humidity. Very high wind speeds, often in mountainous areas can be the reason behind stunted plant growth. Wind also carries seeds and aids in pollination.

Abiotic Components Examples

- Wind.
- Humidity.
- Salinity.
- Rain.
- Temperature.
- Latitude.
- Elevation.
- Radiation.
- Pollution.

Several Types of Ecosystems

The abiotic factors of these ecosystems in the following:

- Desert Abiotic Factors: Due to low rainfall, deserts develop ecosystems which are very distinguishable from another environment. IT covers 20% of the earth's surface and that includes Antarctica. Extreme temperature swings are often observed in deserts as open-air and water vapour stabilizes the temperature.
- Tropical Rainforest Abiotic Factors: Tropical rain forests see the most rainfall on earth. Most rainforests have more than 100 inches of rain every year. It has warm and wet climates and the rainforests create a dense, lush and complex ecosystem.
- Tundra Abiotic Factors: The tundra region receives less light and heat from the sun. A deep layer of soil called the subsoil is observed. And it can remain frozen for many years. Only grasses and small plants grow in this region.
- Ocean Abiotic Factors: The abiotic factors which play a part in the oceans environment are salinity, heat, pollution, and many more. It is a truly unique environment. Because of its depth, the different zones receive a different amount of sunlight and heat. This creates a different ecosystem in each layer and it has its unique share of animals. The different ocean ecosystems are coral reef ecosystems, shoreline ecosystems, deep ocean ecosystems etc.

Other Ranges of Ecosystems

- Temperate Forests: Abiotic factors include, temperature, humidity etc.
- Freshwater Ecosystems: The biotic factors are: light penetration, temperature and pH of water. Examples are lakes, springs.
- Grasslands: As the name says, this type of ecosystem is dominated by grass. The major abiotic factor is rainfall.
- Taiga Ecosystems: It is the coldest region of the arctic. There is a presence of evergreen trees and you can notice mosses and mushrooms.

Responses to Abiotic Factors

Living organisms respond to abiotic components in various ways. This list of abiotic factors includes:

- Regulators: All organisms are able to maintain a constant internal environment called homeostasis. The organisms which are able to do this regulate homeostasis by physiological and behavioural means and it ensures constant body temperature and osmotic concentration. Humans maintain body temperature at 98.4 degrees Fahrenheit. They maintain homeostasis by sweating in the summer and shivering in the winter.
- Conformers: These are organisms which cannot regulate internal body conditions and their body condition changes as per the environment.
- Migrate: For these organisms, when the weather in their habitat transforms into a stressful condition, they move to a habitat which has less stressful conditions. For example, Siberian birds y from that region to Keoladeo National Park in Bharatpur, Rajasthan.
- Suspend: Many organisms have different body mechanisms to survive in a stressful environment. There are many examples of this type of response. These are:
    - Sporulation: Organisms produce thick-walled spores which help the organism to survive unfavorable conditions. When the conditions return to normal, the spore germinates. This type of response can be seen in certain types of bacteria, fungi and lower plants.
    - Dormancy: To survive periods of stress, seeds of higher plants reduce their metabolic activity and go into a state of dormancy. Under favourable conditions, the dormant seed germinates to grow to a new plant.
    - Hibernation and aestivation: If organisms are unable to migrate, they avoid stressful conditions by escaping in time to a place where the organisms sleep in winter. It is called hibernation. If the organism or animal sleeps in summer, it is called aestivation. Example: bears sleep during winter and snails sleep during summer.
- Diapause: It is a natural process which is observed in a certain animal. It causes the delay of development in these animals due to alterations in metabolic activity. Diapause is common in parasites, crabs, shellfish, snail insects and certain groups of zooplanktons.

Biotic Components

Biotic Factors

Biotic factors are the living parts of an ecosystem. Because of the way ecosystems work – as complex systems of competition and cooperation, where the action of every life form can affect all the others – any living thing within an ecosystem can be considered a biotic factor. Biotic factors such as soil bacteria, plant life, top predators, and polluters can all profoundly shape which organisms can live in an ecosystems and what survival strategies they use. Biotic factors, together with non-living abiotic factors such as temperature, sunlight, geography, and chemistry, determine what ecosystems look like and what ecological niches are available.

Types of Biotic Factors

Biotic factors are grouped by scientists into three major groups, which define their role in the flow of energy which all living things in the ecosystem need to survive. These groups are producers or autotrophs, consumers or heterotrophs, and decomposers or detritivores.

Producers

Producers – also known as autotrophs are organisms that make their own food using inorganic materials and energy sources. Producers are extremely important: without them, no life could exist at all. The very first life forms on Earth had to learn to make fuel and building materials to make more cells out of non-living materials. That's because when the first life forms appeared, there were no other life forms to feed on. So the first life forms had to be producers. Producers remain vital today as the life forms that can harness inorganic energy to be used as fuel for life.

There are two major classes of producers:

1) Photoautotrophs are by far the most common type of producer on Earth today. These producers harness energy from sunlight to power their life functions. Green plants, green algae, and some bacteria are photoautotrophs. Most photoautotrophs use a pigment, such as chlorophyll, to catch photons from the Sun and harvest their energy. They then package that energy into a form that all life forms can use, and use it to create proteins, sugars, lipids, and more essential materials for life.

In most ecosystems, plants – which are producers that are multicellular, highly complex, and very efficient at turning sunlight into fuel for living organisms – form the bottom of the energy pyramid. All other organisms depend on the energy plants harvest from the Sun to survive.

2) Chemoautotrophs are fairly rare in most ecosystems. They obtain energy from chemicals such as hydrogen, iron, and sulfur, which are not common in most environments. Nonetheless, they can still play an important role in ecosystems because of their unusual biochemistry. Some methanogens – microorganisms that make methane – are chemoautotrophs. Methane, a greenhouse gas which is much more powerful than carbon dioxide, may play a major role in regulating the planet's temperature. Other chemoautotrophs can produce similarly powerful chemicals with their unique metabolisms.

It is actually not known whether the first forms of life on Earth were photoautotrophs or chemoautotrophs. Photoautotrophs are more common today, but that may simply be because sunlight is more plentiful than the chemicals chemoautotrophs use as their energy source.

Consumers

Consumers, also called "heterotrophs," are organisms that eat other living organisms in order to obtain energy. Herbivores that eat plants, carnivores that eat animals and omnivores who eat both plants and animals, are all heterotrophs. Heterotrophy probably evolved when some organisms discovered that they could eat autotrophs as a source of energy, instead of creating their own energy and organic materials.

Some autotrophs subsequently evolved symbiotic relationships with consumers, such as angiosperms – plants which produce nectars and fruits to attract animals, who ultimately help them to reproduce. Most levels of most ecosystems' energy pyramids consist of consumers – herbivores, minor predators, and top predators who eat other organisms.

Decomposers

Decomposers, or detritovores, are organisms that use organic compounds from producers and consumers as their source of energy. They are important to ecosystems because they break down materials from other living things into simpler forms, which can then be used again by other organisms. Decomposers include soil bacteria, fungi, worms, flies, and other organisms that break down dead materials or waste products from other life forms. They are distinct from consumers, because consumers usually consume other organisms while they are still alive.

Decomposers, on the other hand, metabolize waste products that might not be of interest to consumers, such as rotting fruit and dead animals. In the process they break down these dead things into simpler chemicals that can be used by heterotrophs to thrive and produce more energy for the ecosystem as a whole.

This is the principle behind the practice of composting – where waste scraps of plants and animal products are put into a pile, where decomposers such as bacteria, worms, and flies are allowed to thrive. These decomposers turn the waste products into rich fertilizer for the composter's garden, which then grows bigger and healthier thanks to the decomposers breaking down the waste products in the compost.

Decomposers are the link between the bottom of an ecosystem's energy pyramid and the other levels. Decomposers can take energy and raw materials from dead plants, herbivores, lesser carnivores, and even top carnivores, and break it down into a form that can be used by the ecosystem's producers to make it easier for them to harness sunlight. In this way, the ecosystem's energy cycle is preserved.

Examples of Biotic Factors

Cyanobacteria and Life on Earth

Scientists believe that the earliest widespread form of life on Earth was cyanobacteria. These fairly simple cells, which made food and organic materials from sunlight, played a massively important role in creating all of Earth's modern ecosystems. Prior to the success of cyanobacteria, Earth did not have an oxygen atmosphere. That meant that aerobic respiration was not possible – and also meant that it was impossible, or very difficult, for any organisms to live on land because of the DNA-destroying ultraviolet radiation from our sun.

However, cyanobacteria developed a method for storing the energy of sunlight in organic molecules. For this they needed to take molecules of carbon from inorganic sources, such as carbon dioxide in the air, and turn them into carbon-based organic compounds such as sugars, proteins, and lipids.

To achieve this, cyanobacteria took in the inorganic gas CO_2, and released a new gas, O_2. O_2, or molecular oxygen turned out to be the perfect fuel for the most powerful type of heterotroph metabolism: aerobic respiration. Molecules of O_2 also reacted with ultraviolet light in the upper atmosphere to form, O_3 – a molecule also known as ozone, which absorbed ultraviolet light in the upper atmosphere and made it safe for life forms to colonize land.

In the billions of years to come, cyanobacteria would be mostly replaced by its more sophisticated descendants such as trees, grass, and algaes who would take over its role as Earth's primary oxygen producers. However, cyanobacteria itself still appears in blooms which can sometimes be seen from space. As biotic factors, cyanobacteria and its modern descendants supplied not only energy and organic compounds, but also oxygen, to all of Earth's ecosystems.

Filamentous Cyanobacteria Bloom.

Wolves in North America

When European colonists arrived in North America, wolves were common in many of the continent's ecosystems. These large carnivores were the top predators in many places, using a combination of their large size and teamwork to take down large prey animals. The colonists and their descendant's hunted wolves fiercely, due to safety concerns over the fact that wolves could eat sheep that farmers depended on for food, and could even eat human children.

However, the disappearance of wolves eventually started to cause new problems for the humans of North America. Without their top predator, deer and other herbivore species multiplied to unprecedented numbers.

This might have seemed nice for human hunters who ate deer meat and sold deer skins at first, but the problem became serious when the deer started eating so many plants that crops, gardens, and wild plant species became endangered. Humans began to have to hunt deer themselves, not just for meat and skin, but to prevent serious damage to their ecosystems.

Humans didn't realize the full extent of the roll of wolves until a ban on wolf hunting was introduced, and wolves bred in captivity were released back into the wild to repopulate the wolf species in some areas. The areas where wolves were re-introduced underwent startling transformations. Numbers of deer and other large prey species went down, sure enough – which led to populations of many plant species increasing.

To the surprise of human scientists studying the ecosystems, even land forms began to change: it turned out that deer had been eating grass and other small plants whose root systems held soil in place against erosion. With wolves keeping the deer population in check, the plant populations began to come back – and erosion decreased, and the courses of rivers changed. Fish were also affected by a decrease in loose soil washing into the river.

This is an excellent example of how complex and interconnected ecosystems are – and how the removal of one element of the ecosystem, even if its only role is to eat other animals, can cause big changes for all other organisms living in the ecosystem.

Humans

In 2016, biologists around the world decided to declare that the Earth had entered a new geologic era: the Anthropocene. This era is defined by the effects of human technology, which has caused massive changes to the global ecosystem on par with the effects of past major climate change events and even asteroid impacts.

Human activity has drastically changed the carbon cycle of Earth, with the burning of wood, coal, and oil releasing millions of years' worth of carbon dioxide into the atmosphere in the space of just a couple of centuries. Over the same time scale, humans have cut down about half of all Earth's forests, which had previously acted to take carbon dioxide out of the air and incorporate it back into plant life.

In addition, humans have begun to release many new substances into Earth's land, air, and oceans, including plastics, heavy metals, and radioactive materials, none of which exist in nature. The result has been the beginning of alarmingly rapid climate change and a mass extinction, in which species are disappearing faster than they have been since the asteroid impact that killed the dinosaurs and made way for the rise of mammals 65 million years ago.

Humans are therefore maybe the most powerful example of how the living factors in an ecosystem can change it since cyanobacteria. This has led some environmentalists to suggest that humans are "evil," and "bad for the Earth. But the truth is that the Earth always survives ecological upheavals. It's just a question of whether the species that exist at their beginning survive to their end.

That's why many scientists say humans should be concerned about their effect on the planet. Not because changing the planet is in itself morally wrong, but because humans themselves rely on the complicated ecological interplay of thousands of species for their food. Scientists are already beginning to raise alarms that the pollinators on which many human food crops depend appear to be dying off due to new chemicals that humans have released to the environment.

Human food crops are also threatened by climate change caused by the carbon dioxide humans have released into the air, which has brought severe drought to many areas with dense human populations who require large amounts of food to survive.

Medical scientists also caution that man-made climate change is allowing dangerous insect-borne diseases, which used to be restricted to the regions near the equator, to spread to new areas. As the dominant species on Earth, it is important that humans learn about the ecosystems upon which they depend for well-being and survival.

Ecological Niche

Ecological niche is a term for the position of a species within an ecosystem, describing both the range of conditions necessary for persistence of the species, and its ecological role in the ecosystem. Ecological niche subsumes all of the interactions between a species and the biotic and abiotic environment, and thus represents a very basic and fundamental ecological concept. The tentative definition presented above indicates that the concept of niche has two sides which are not so tightly related: one concern the effects environment has on a species, the other the effects a species has on the environment. In most of ecological thinking, however, both meanings are implicitly or explicitly mixed. The reason is that ecology is about interactions between organisms, and if persistence of a species is determined by the presence of other species (food sources, competitors, predators, etc.), all species are naturally both affected by environment, and at the same time affect the environment for other species.

If we want to treat both of these aspects of ecological niche within one framework, we can define it more formally as the part of ecological space (defined by all combinations of biotic and abiotic environmental conditions) where the species population can persist and thus utilize resources and impact its environment. It is useful, however, to distinguish three main approaches to the niche. The first approach emphasizes environmental conditions necessary for a species presence and maintenance of its population, the second approach stresses the functional role of species within ecosystems, and the third one a dynamic position of species within a local community, shaped by species' biotic and abiotic requirements and by coexistence with other species.

Concepts of Niche

Niche as the Description of a Species' Habitat Requirements

The first formulations of the concept of an ecological niche were close to the general meaning of the term: the ecological niche was defined by a place a species can take in nature, determined by its abiotic requirements, food preferences, microhabitat characteristics (for example a foliage layer), diurnal and seasonal specialization, or predation avoidance. This concept is associated mostly with Joseph Grinnell, who first introduced the term. He was especially interested in factors determining where we can find a given species and how niches, generated by the environment, are filled. The knowledge of a species niche determined by its habitat requirements is essential for understanding and even predicting its geographic distribution; this concept of the niche is thus more relevant in biogeography and macro ecology than in community or ecosystem ecology.

Niche as Ecological Function of the Species

In this concept of a niche, each species has a particular role in an ecosystem and its dynamics, and one such role can be fulfilled by different species in different places. The observation of distant species adapted to equivalent ecological roles (the resemblance between jerboa and kangaroo rat, between many eutherian and marsupial species, or the Galapagos finches diversifying to highly specialized roles including those normally taken by woodpeckers) was clearly influential to Charles Elton, who emphasized the functional roles of species. According to Elton, there is the niche of burrowing detritivores, the niche of animals specializing in cleaning ticks or other parasites, or the pollination niche. Elton's niche can apply to several species, for example "the niche filled by birds of prey which eat small mammals". This functional niche therefore refers to a species position in trophic chains and food webs, and the concept is thus especially relevant for ecosystem ecology.

Niche as a Species Position in a Community – Formalization of Ecological Niche Concept

The emphasis on the diversity of ecological communities and interspecific competition within them in the second half of 20th century has led to the formalization of niche concept, and an emphasis on the properties of the niches which enable species coexistence within a habitat. George Evelyn Hutchinson postulated that niche is a "hypervolume" in multidimensional ecological space, determined by species requirements to reproduce and survive. Each dimension in the niche space represents an environmental variable potentially or actually important for a species' persistence. These variables are both abiotic and biotic, and can be represented by simple physical attributes as temperature, light intensity or humidity, but also more sophisticated attributes such as soil texture, ruggedness of the terrain, vegetation complexity or various measures of resource characteristics. This could be viewed simply as a formalization of original Grinnellian niche, i.e. the exact descriptions of species habitat requirements. However, in the Hutchinsonian view ecological niches are dynamic, as the presence of one species constrains the presence of another species by interspecific competition, modifying the position of species' niches within the multidimensional space. This concept therefore combines the ecological requirements of the species with its functional role in the local community.

Fundamental and Realized Niche

Hutchinson recognizes a species' fundamental niche, a multidimensional 'cloud' of favourable conditions determined by all environmental (abiotic and biotic) variables where the species can reproduce and survive, and the realized niche, which is a subset of the abstract fundamental niche, where the species can persist given the presence of other species competing for the same resources.

Thus, a realized niche always has a narrower extent along respective dimensions; a species which could potentially live in a broad range of humidity conditions, for instance, may occupy a much narrower range of these conditions in an environment with competing species, since its population growth rate decreases to negative values in some conditions. To a good approximation, if we ignore stochastic sampling from a heterogeneous species' population, species does fill its realized niche.

Niches of different species can be separated along any of these dimensions or by a combination of them (i.e. their interaction). Although this formal model of the niche has quite straightforward theoretical consequences, in practice it can be quite difficult to describe properly the ecological niches of real species, because the number of niche dimensions is potentially infinite, and the significant niche axes (and appropriate measurements) may be rather hard to find; a niche overlap among species may mean we did not succeed in determining the crucial niche axes of separation. However, often a few variables are sufficient to separate species' realized niches, and they or their correlates can be inferred assuming we understand the species' biology reasonably well. For example, five species of warblers, analysed by Robert MacArthur, showed significant (though incomplete) separation along only three niche axes (feeding behaviour, feeding height and nesting time).

The difficulties in determining appropriate niche axes, however, still considerably limit the usefulness of the concept in empirical research. Even if we know the important resources, it is still problematic to decide which characteristics to measure. A further problem, albeit rather technical, is posed by including discrete categories: the width of the cloud in the respective dimension would be reduced to zero, and its position can be arbitrary. More importantly, although species can often potentially live in a much broader range of environmental conditions than where they do actually live, the distinction between the "fundamental" and "realized" niche is slightly arbitrary, driven purely by the interest in coexistence of species sharing resources. As the dimensions of the fundamental niche are both abiotic and biotic, there is no a priori reason to exclude presence of competitors from the dimensions characterizing fundamental niche. The distinction between the fundamental and realized niche may also be blurred, as species' interactions need not fit to our discrete categories - for example, competitors may act also as predators.

Due to the difficulties with the concept, and for tractability, a considerable part of the theory actually dealing with species coexistence works with a one-dimensional approximation of the trophic niche, a resource utilization function – given by the frequency distribution of an important characteristic of utilized resource (for example, a prey size).

Competitive Exclusion

Historical development of the niche theory is very closely related to one of the most important topics of ecology, i.e. the problem of species competition and coexistence. Since the beginning of the ecological niche concept, it has been assumed that no two species sharing a single niche could locally coexist. Originally, the later Volterra-Gause principle states that "under constant conditions, no two species utilizing, and limited by, a single resource can coexist in a limited system" and was formulated and proved by Vito Volterra, whilst Alfred Gause showed experimental evidence of competitive exclusion in an undiversified environment. The explanation of the competitive exclusion lies in the fact that utilization of a limited resource leads to its depletion, and the population growth, therefore, necessarily leads to a moment when the resource level is insufficient for further growth. If only one population utilises the resource, this situation leads to simple negative feedback, causing the decrease of population growth rate and thus a release of resource consumption, stabilizing the population size. However, in the case of two species sharing the resource, there will likely exist a resource level when the first species can still grow up even if the second cannot, leading to further decrease of population growth rate of the second species, and eventually to its extinction. Even if two species sharing several resources have exactly the same requirements and ability to utilize them, the coexistence of such species is not stable due to environmental or demographic stochasticity: over time, one of the species would ultimately become extinct by chance.

The "competitive exclusion principle" is the core principle in community ecology, and much of this field has been devoted to study how species with similar ecological requirements can coexist. This question has transformed into the problem of "limiting similarity": how similar can ecological niches be to still ensure local coexistence.

Limiting Similarity, Species Packing – How Close can Species be to Each Other?

Hutchinson states that a species' realized niche is exclusive, i.e. no two species can share a single niche and no overlap in the realized niches is possible in a stable environment. In other words, were there to be an overlap in, say, the trophic "dimension" of the niche, species would differ in other dimensions - for example in their tolerance to abiotic factors, or avoidance of predators. Now, the (rather vague) consensus is that a little overlap between niches is consistent with coexistence, whereas somewhat larger overlap is not. The theory of limiting similarity, formalized by Robert MacArthur and Richard Levins, predicts the minimum permissible degree of overlap in the resource utilization curve. They showed that coexistence between species utilizing a continuous resource is possible when the ratio between the niche width and the distance between species' optima is approximately unity or smaller. (This has been derived using the Lotka-Volterra equations describing the growth rates and hence stability of populations of competing species, where the competition coefficients were determined by the proximity of species' bell-shaped utilization curves). However, the result is sensitive to the assumptions about the form of the resource utilization function and population growth rate; notably, highly peaked resource utilization functions show actually almost no limits to coexistence (as their overlap is always minute) and niches can overlap broadly when fitness increases as the frequency of individuals carrying the respective trait decreases – under negative frequency-dependence. Also, coexistence between species can be facilitated by nonlinear responses of the competitors to common fluctuations in the environment.

The predictions of the theory of limiting similarity cannot be directly corroborated by observation: by definition, the population density of one of the species is close to zero if the species pair is close to limiting similarity, and thus the utilization functions are not observable in such a situation. On the other hand, finding a similarity higher than predicted would clearly indicate that some of the assumptions of the model are violated.

The spacing between species in niche space, resulting from partitioning the available resources (species packing), differs considerably between sexually and asexually reproducing species. In sexually reproducing populations, recombination generates all possible combinations from the common gene pool, although some of these may be maladaptive. Tradeoffs in utilizing the resource spectrum may – or may not generate disruptive selection strong enough to drive evolution of reproductive isolation and evolution of distinct species, thus eliminating maladapted recombinant genotypes. In contrast, in asexual species, clones bearing favourable combinations do not recombine, and therefore those adapted to the various resource combinations can be arbitrarily spaced in the niche space. Due to the necessity of finding a mating partner, population growth rate of sexual populations can sharply decrease at low densities (Allee effect), limiting both adaptation to marginal conditions and invasion to a new area. Both these effects contribute to discontinuities in distribution of resource use of sexually reproducing species.

Modes of Species Coexistence

Species coexistence is often ensured by niche separation. The niche shift can follow from the competitive exclusion of one species from the part of ecological space where the niches overlap, or from coevolution of competing species, favouring in each species phenotypes differing from the phenotype of the competitor. Current niche segregation can be due to the processes that took place in distant evolutionary past – sometimes emphasized by the term "the ghost of competition past". When phenotypic differences arose due to divergent evolution of sympatric competitors, we speak of character displacement. Typically, sympatric populations of competing species evolve towards more different sizes of characters associated with food consumption (beaks, teeth) than allopatric populations – if there is an island with only one species of Galapagos finches, it has an intermediate beak size enabling it to utilize a wide spectrum of seed sizes, whereas if there is an island with two species, one has a bigger and the other a smaller beak than the species occurring without competitors. If there are more than two locally coexisting species, we often observe regularly spaced sizes of morphological characters, again indicating past competition leading to maximum niche separation.

Simple separation of niche optima is not, however, the only way that stable local coexistence of species is attained. Many species pairs, for instance, consist of one species which is competitively dominant and the other species which is less specialized and can thrive in a broader range of ecological conditions. An example is the pair of two closely related species of redstarts, where the black redstart *Phoenicurus ochruros* is bigger and more aggressive, but the common redstart *Phoenicurus phoenicurus* can utilize a wider spectrum of habitats, such that it always has an option to thrive outside of the range of conditions preferred by the black redstart. Such niche division between dominant aggressive specialist and subordinate generalist has also been observed in many mammal species, and is apparently stable. In plants, competitively inferior species are often those with higher rates of spreading and growth, which enable them to quickly occupy empty places before arrival and eventual overgrowth of competitively superior species. In this case, we speak about the regenerative niche, representing a time window for the success of competitively inferior, but fast spreading and fast growing species, thus ensuring long-term coexistence of competitors in the same habitat.

If species are very similar to each other, such that they do not differ substantially in their utilization of resources, the competitive exclusion can take a very long time. If the replacement of old individuals by young ones is basically a random process, i.e. all individuals regardless of species identity have equal chances to give birth to their descendants within an environment, populations of all involved species will fluctuate randomly and the prevalence of a particular species is just a matter of chance. However, due to these stochastic fluctuations and due to the fact that the species which incidentally prevails in a time step will have higher probability to further increase its abundance, this process will finally lead to apparent competitive exclusion. This process, called community drift, can be relatively slow and may be further slowed down by dispersal limitations (leading to random prevalence of different species in different local communities isolated by migration barriers) and balanced by the emergence of new species (i.e. speciation or migration from elsewhere).

Communities where dispersal limitation and community drift play a major role are called dispersal-assembled communities, in contrast to niche-assembled communities where niche differences play a major role in determining species distributions and abundances. Trees in tropical forests represent a very good candidate for dispersal-assembled communities. Most tropical tree species are very similar in terms of their ecology and growth characteristics, and it has been documented that for their recruitment the proportion of parent individuals in a given locality (i.e. dispersal limitation of more distant individuals) is much more important than any habitat characteristics. Still, an incredible number of species can coexist locally. It is hardly believable that there are several hundreds of different narrow ecological niches (i.e. combinations of environmental characteristics) on a hectare of tropical forest to enable coexistence of several hundred tree species on the basis of their niche differences – the dispersal assembly and coexistence without significant niche differentiation seems more likely.

However, an unusual aspect of niche differences can still be involved in this classic case of species coexistence. It has been demonstrated that coexistence of tropical trees is facilitated by frequency-dependence, where relatively rare species have an advantage of not being so severely attacked by natural enemies which strongly limit recruitment of more common species on which they specialize. In a sense, all species compete for "enemy-free space", and this "niche" for a given species is open only if the species is not too abundant to allow population growth of specialized natural enemies. Separation of "niches" of tropical trees seems thus to be determined by the community of species-specific pathogens.

Coexistence among species can be certainly maintained both by niche differences and – at least in a non-equilibrium world – by niche similarity. Coexistence of species with similar niches maintained by dispersal-assembly processes could be a reason why we often observe that species are not regularly distributed in a niche space, but form clumps of species whose niches are closer to each other than to other species.

Types of Ecological Niche

The notion that ecological niches cannot be infinitely similar to each other, and the knowledge that ecological space is heterogeneous and that the distribution of resources available to a community is always limited, has led to an idea that for a given environment there is a limited number of available niches which could be potentially occupied. An environment then could be seen as a set of empty niches, which could – but may not – be filled with species. Consequently, we might ask whether in a particular case the niche space is or is not saturated with species.

There are two facets of the problem, which are sometimes confused. First, there is no doubt that the limited amount of resources in an ecosystem can sustain only limited total number of individuals (assuming a given body size distribution). Therefore, there is always a limited potential for the whole community size determined by total amount of resources, and thus also for a limited number of species (given that each species needs some viable population size). If this potential is fully utilized, we speak about biotic saturation of the community.

However, biotic saturation does not imply that the number of ecological niches is fixed and that all possible niches are occupied. Such a statement would be much stronger and would require at least some level of discreteness of ecological niches, i.e. that ecological space cannot be divided into an infinite number of subtly different niches with arbitrary positions. Is there any reason to believe that niches are discrete and their number within an environment is limited?

Apparently, there is a considerable level of environmental heterogeneity in resource distribution and abundance; resources are more abundant for some combinations of parameters than for other. Environmental heterogeneity would not be, however, a sufficient condition for discreteness of ecological niches if species could utilize equally easily several different resources. The discreteness of ecological niches comes out from the existence of trade-offs in resource utilization: resources can always potentially be utilized in many ways, but a species which utilizes many resources typically does so with a lower efficiency compared to a specialist. A Galapagos finch from the genus Geospiza can have either a big beak appropriate for cracking big seeds, but then it can crack small seeds with much more difficulty – and vice versa. Some strategies in resource utilization are mutually exclusive: a plant can either invest to its rapid growth or so quickly utilize resources, or it can invest into woody trunk which enables it to grow higher and sustain longer – but at a cost associated with slower growth. Whilst natural selection supports phenotypes that are better at utilizing available resources, it can only explore the vicinity of the current strategy, selecting from the available phenotypes. Consequently, evolution leads to utilization of only a restricted spectrum of resources.

In the presence of trade-offs, there is only a limited number of mutually exclusive ways to utilize resources, and thus a limited number of available niches. However, as the discreteness of niches follows from the trade-offs between adaptations, and since all the trade-offs are determined by unique properties and constraints of given organisms, it makes sense to speak about available niches only in relation to organisms which already inhabit the environment. A habitat without its inhabitants can provide a potentially infinite number of opportunities for existence, and this landscape of opportunities changes with each new inhabitant. For the organism in an environment, the number of possible niches is determined by the number of possible ways to utilize the resource - with all constraints and trade-offs of the given organism. Therefore, it is likely that there are always more niches than the current number of species, because each species has several mutually exclusive possibilities of future adaptive evolution arising from the trade-offs unless all niche changes require a corresponding niche change in other species.

In some cases, the number of available niches can be predicted from the knowledge of resource heterogeneity and the possibilities of resource utilization for given taxon. The number of Galapagos finches occurring on each island is reasonably well predicted by the number of peaks of the "landscape" constructed using the knowledge of frequency distribution of seed size, the general relation between finch and seed biomass, and the relation between preferred seed size and beak depth. Similarly, using the knowledge of the relationship between beak shape of crossbills (Loxia curvirostra) and their foraging efficiency in obtaining cone seeds from cones of various coniferous tree species, it is possible to construct a resource utilization function related to different morphologies, and find out how many optimal shapes do exist. And again, it has been found that there are several ecomorphs of crossbills, each of them occupying one adaptive peak (optimum) in the morphological space.

There is other evidence that ecological niches are partially predictable – the phenomenon of community convergence. Animal or plant communities occurring on different continents or biotic provinces often comprise similar morphological types utilizing similar types of resources. Anolis lizards, for instance, have evolved independently into several well- recognizable ecomorphs on each Caribbean island, with known sequence of this evolution, repeated on every island. However, there can be more than one species within each ecomorph, and thus this convergence does not imply that the number of species-specific niches is predictable. This is quite typical for most cases of community convergences: they provide a clue to our understanding of how many possibilities are there for utilizing resources within a given habitat and for a given taxon, but not to the prediction of how many species can actually coexist there. The total potential number of species within an environment is given by the total amount of resources determining the total number of all individuals, regardless the level of discreteness of ecological niches.

Ecological Niches and Patterns in Species Abundance and Distribution

Species spatial distributions as well as their abundances are often attributed to the breadth and position of their niches. A species occurs in places where its requirements are fulfilled, i.e. where it finds its niche. However, the "presence of the niche" is not a sufficient condition for the presence of a species, and in special cases it may not be even the necessary condition.

Spatial population dynamics driven by dispersal and spatial distribution of available habitat patches is equally important. Consequently, species may be absent even in sites containing habitat that fulfills its niche requirements if the site is far away from other occupied sites and the dispersal distance of the organism in concern is relatively small, hindering immigration into the site. On the other hand, a species may be present even in a site where its niche requirements are not fulfilled and population growth is negative if the population is maintained by a continuous supply of individuals from neighbouring sites with positive population growth (so-called source-sink population dynamics). Therefore, species spatial distributions are determined by species niches and available habitat distributions, as well as by spatial population dynamics and dispersal limitation.

In a similar line, it has been argued that a significant proportion of the variation of species' abundances can be explained by the breadth of species' niches. It is reasonable to assume that species which are able to utilize wider spectrum of resources can attain higher population abundances and also can occupy more sites. Local population densities are mostly positively correlated with species range sizes, which can be taken as evidence of such niche differences. However, patterns in species abundances can be often well explained by spatial population dynamics – for instance species which were incidentally able to spread to more sites have higher chance to colonize further sites and to further increase local population densities by immigration (this is the non-linearity of the dynamics of met populations). Moreover, the statistical relationship between niche breadth and abundance can actually have a reversed causality, as abundant species are forced to utilize a wider range of resources due to intraspecific competition. More abundant species can also be those that do not utilize a broader range of resources, but are specialized on resources which are relatively more abundant, or may simply have higher population growth and/or dispersal rate (although these features can be understood as niche properties).

One of the most prominent ecological patterns is the frequency distribution of abundance of individual species within local communities or regional species assemblages - the so called species-abundance distribution. It is always highly unequal, the majority of species having low abundance and only a few being common (the frequency distribution is often close to lognormal, though other models may fit the observed species-abundance distribution better in particular situations). This distribution has been modeled as a stepwise division of niche space, where each newly arriving species obtains some (random) proportion of niche space previously utilized by other species. One of these models, based on sequential resource partitioning, predicts observed species-abundance distribution quite well. However, models based on spatial dynamics and dispersal limitations – especially those involving "community drift" can provide equally good predictions of species-abundance distribution. This again indicates the complementarity between niche-based and dispersal-based explanation of ecological patterns, and supports our consideration of both niche differences and spatial population dynamics as essential drivers of species distribution and abundance.

Niche Divergence and Resource Specialization

The diversity of ecological niches even among closely related species is enormous and demands explanation. More specifically, natural selection prefers such phenotypes of competing species which utilize different resources than those which share them. Competition thus leads to the increase of the resource range utilized by a given taxon, and this process is faster when other taxa do not constrain this diversification. Indeed, the increase of the breadth of utilized resources in the course of evolution is fastest in such situations where other taxa with similar requirements are absent. For example, Ecomorphological diversification of Galapagos finches and Hawaiian honeycreepers has been much faster than the diversification of related taxa on the mainland, where the utilization of new resources was constrained by other taxa already utilizing them.

Interspecific competition is not, however, the only force driving niche diversification. Each species has its own evolutionary history, and thus can adapt to different resources by an independent process of evolutionary optimization, as phenotypes which are more efficient in transforming obtained energy into offspring are favoured by natural selection. If there are several mutually exclusive ways to achieve this, it is likely that each species will go by a different route due to evolutionary contingency, and niche diversification will follow without competition. Notably, optimization does not lead to an advantage of the whole species in terms of the resource utilization but only to an individual advantage regardless of the evolutionary fate of the whole species: when the niche becomes narrower, the species' range and hence total population size decreases. As evolution is opportunistic, species can evolve to extremely specialized forms in terms of either habitat utilization or food preference, which is apparently disadvantageous for future species persistence in ever changing world.

Evolution of Niche Width

Progressive specialization, i.e. narrowing of niche width in the course of evolution, is forced by interspecific competition (when the niches overlap) and intraspecific optimization, and thus represents an expected evolutionary trend. The opposite process, i.e. an extension of niche width, is observed mostly after entering a new environment without competitors, allowing utilizaton of a wider spectrum of resources. This process is called ecological release and may be underlined both by the extension in within and between phenotype component of species ecological variation (importance of the two contributing modes vary widely among species).

Species niches can widen also because of phenotypic plasticity (heritable genotype environment interactions directing the trait in the early ontogenesis), and can vary even purely behaviourally, as an immediate response to an altered resource or species structure.

Although sometimes there is an obvious constraint on expanding a species' niche - e.g. physiological constraints like freezing of body fluids or presence of a competing species - we often see no apparent reason why species niches stay restricted to a fraction of a resource which continuously varies in space. However, even a gradual change in the environment can lead to a sharp range margin, as "swamping" by gene flow from central to marginal areas can create a positive feedback between the reduction in fitness, where population size decreases due to maladaptation, and the erosion of genetic variance by genetic drift. When local genetic variance is too low, continuous adaptation to a spatially variable environment fails. Moreover, when there is a large asymmetry in the carrying capacity across the habitats, even alleles which neutral or nearly neutral in the main population and deleterious in the marginal populations, can sweep through the small marginal population, thus preventing the adaptation.

Asexual reproduction or self-fertilization can hence provide an advantage in adapting to marginal conditions – both because small populations are still viable (as there is no need to find a mating partner) and because gene flow does not restrict adaptation to marginal conditions. Indeed, it is found in many plants and animals adapting to extreme, marginal habitats (classic animal examples are Daphnia pulex or freshwater snail Campeloma). However, although lack of recombination in asexuals means that locally favourable gene combinations are maintained, adaptive evolution in asexual species is significantly slowed down as beneficial combinations have to arise in each strain independently. It appears that high levels of, but not obligatory, self-fertilization or asexual reproduction (parthenogenesis and vegetative reproduction) are commonly advantageous for adaptation to marginal habitats.

In some cases, we observe an apparent regularity in the evolution of niche width and position. The classical example represents cycles of species dispersal, specialization and local adaptation (and eventual extinction) observed on various archipelagoes, called taxon cycles. They were originally described by Edward O. Wilson on Melanesian ants, but were best documented by Robert Ricklefs and his coworkers on Caribbean birds. In the first stage, an immigrant, which is mostly a species with high dispersal ability, colonizes coastal or disturbed areas. Then the species spreads across the island, adapting to the new resources and expanding its niche (quite likely as a consequence of a release from competitors, predators and parasites). In the next step the species becomes more specialized, and its distribution becomes spottier. The narrowing of its niche may be driven by an immediate advantage of an adaptation to a local resource or immigration of new generalist competitor. Finally, species distribution becomes very fragmented, ending in local endemism, and ultimately extinction.

Changing the Niche Space, Niche Construction and Coevolution

Both environment and species change in the course of time, and thus ecological niches are not stable and given forever. Species not only respond to environmental changes, but also actively change their biotic and abiotic environment, affecting both their own niche and the niches of other organisms. The importance of competition, predation or mutualism has been already stressed. Moreover, organisms often make niche space for other organisms available in the environment – think of the resource space generated by an emerging tree or of successive colonization of an island, where first colonizers modify the environment for their successors.

Some organisms strongly directly affect abiotic environment, determining possible niches for a whole community of species. Beavers building dams, earthworms altering soil structure or, on a larger scale, plants providing oxygen are classical examples. Organisms substantially affecting abiotic environment are often called ecosystem engineers and the process in which an organism systematically modifies its own niche (both biotic and abiotic components), is called niche construction. Obviously, this process is most pronounced in Homo sapiens, which is currently the most conspicuous ecosystem engineer.

If a species can change its environment as well as to adapt to it, coevolution between a species and its niche can follow, based on the continuous feedback between a species' niche construction and its adaptation. Since species continuously change the environment for themselves as well as for other species, species' niches can be very dynamic. Often, however, species' ecological requirements are quite stable over evolutionary time, so that it is even possible to reliably reconstruct an ancient environment on the basis of presence of particular species in the fossil record and the knowledge of their contemporary ecological niches. This can be attributed to the fact that it is often easier to search for appropriate habitat elsewhere if an environment within a locality is changing than to adapt to it. All species have some dispersal abilities, and are thus able to track spatio-temporally changing habitat availability by migration rather than adapt to different conditions by mutation-selection process. The other reason for the apparent niche conservatism is the existence of evolutionary constraints and consequent trade-offs: species often cannot easily change their traits in a particular direction if these traits are associated with other traits whose change is not advantageous.

As species undertake evolutionary changes, their functional niches can change, leading to changes in the overall "ecological space" in an ecosystem, and promoting further changes in species traits. On the other hand, some functional niches, i.e. particular ecological roles, can be rather stable even if species evolve, go extinct and new species emerge - similar functional role can be progressively fulfilled by different species. Community evolution can be therefore viewed as a coevolution of ecological niches rather than of species themselves.

Ecological Succession

Ecological Succession is also called as Plant Succession or Biotic succession. Hult used the term "succession." The occurrence of relatively definite sequence of communities over a long period of time in the same area resulting in establishment of stable community is called ecological succession. It allows new areas to be colonized and damaged ecosystems to be recolonized, so organisms can adapt to the changes in the environment and continue to survive.

Major Types of Ecological Succession

- Primary Succession: When the succession starts from barren area such as bare rock or open water. It is called primary succession.

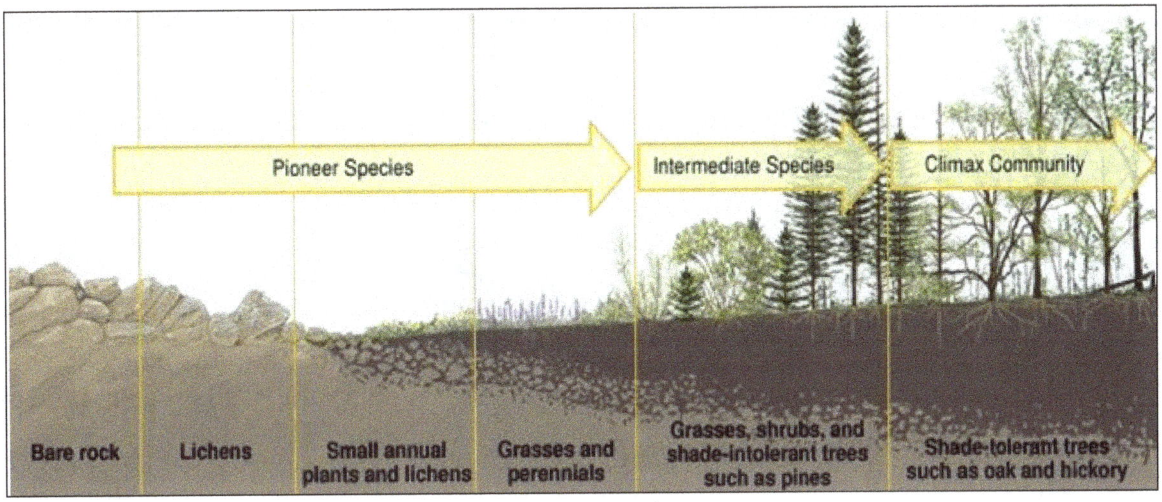

Primary succession.

- Secondary Succession: Secondary succession occurs when the primary ecosystem gets destroyed by fire or any other agent. It gets recolonized after the destruction. This is known as secondary ecological succession.

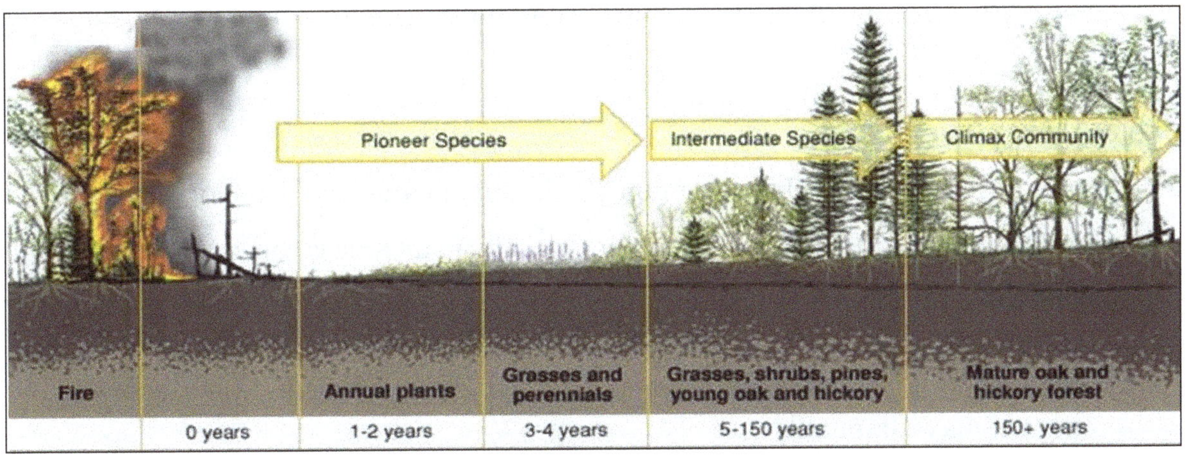

Secondary succession.

- Autogenic succession: After succession has begun, its vegetation itself causes its own replacement by new communities is called Autogenic succession.
- Cyclic Succession: This refers to repeated occurrence of certain stages of succession.
- Allogenic succession: When the replacement of existing community is caused by any other external condition and not by existing vegetation itself. This is called allogeneic succession.
- Autotropoic succession: It is characterised by early and continued dominance of autotrophic organism called green plants.
- Heterotropic succession: It is characterised by early dominance of heterotrophs such as bacteria, actinomycetes, fungi and animal.
- Induced succession: Activities such as overgrazing, pollution and scarping may cause induced succession.

- Retrogressive succession: A return to simpler and less dense community is known as retrogressive succession.
- Directional succession: The progression of succession on the overall community level is called directional succession.

Process of Succession

Clements has discussed the following process of succession:

- Nudation: Development of bare area without any form of life. The cause may be climatic (glacier, hails), biotic (epidemic, human activities) or topographic (soil erosion, landslide).
- Invasion: Successful establishment of a species in a bare area. May be by migration, ecosis or aggregation.
- Competition: It involves the development of intraspecific as well as interspecific members. Competition occurs due to limited food and space.
- Reaction: It involves the modification of the environment through influence of living organism.
- Stabilization or climax: The final terminal community becomes more or less stabilized for a longer period of time is climax community.

Types of Communities

- Pioneer community: The first community to inhibit an area is called Pioneer community.
- Climax community: The last and stable community in an area is called Climax community.
    - Pro-climax is referred to that community which is more or less stable resembling the climax community in some respects.
    - Sub-climax is a stage or community in an ecological succession immediately preceding a climax community.
    - Post-climax, is the new climax formed.
    - There are three different theories regarding the concept of climax:
        - Monoclimax theory: Proposed by Fredrick Clements in 1916. According to this theory there is only one climax community in a particular climatic region.
        - Polyclimax theory: Proposed by Tansley in 1935. According to this theory a number of different climax communities may be present in a particular climatic region.
        - Climax pattern hypothesis: Proposed by R.H. Whittaker in 1953. According to this hypothesis there is one big community that changes according to soil, slope and other habitat factors.
- Seral Community/Transitional Community: A seral community is an intermediate stage of ecological succession advancing towards the climax community. A seral community is replaced by the subsequent community. It consists of simple food webs and food chains. The entire series of communities is called Sere. Individual transitional communities are called Sere stage.

Types of Seres

Types of Seres	Explanation
Hydrosere	Succession in aquatic habitat.
Xerosere	Succession in dry habitat.
Lithosere	Succession on a bare rock surface.
Psammosere	Succession initiating on sandy areas.
Halosere	Succession starting in saline soil or water.
Senile	Succession of microorganism on dead matter.
Eosere	Development of vegetation in an era.

Hydrosere

It is succession occurring in the aquatic environment. If the body of water is large and very deep or very strong wave action and other powerful physical forces are at work, the succession change is hardly recognizable. Succession is recognizable only if the colonization of plant communities takes place in artificial small and shallow ponds, lakes, etc. where wave action speeds up the process by allowing the erosion of soil towards edge regions. In this way, the filling process also speeds up quickly and consequently the body of water disappears within few years' time. Various stages of succession in a lake or pond are:

- Plankton stage: The pioneer community starts with germination of spores which reach in water through wind or animals. Phytoplanktons =>Diatoms, phytoflagellates, Cynobacteris, Green algal cells. Zooplanktons are formed due to phytoplanktons. Death and decomposition of organisms add large amount of organic matter and nutrients.
- Submerged stage: When a loose layer of mud is formed on the bottom of the pond, some rooted submerged hydrophytes begin to appear on the new substratum. The submerged aquatic vegetation develops in the regions of ponds or lakes where water depth is about 10 feet or more. The pioneers are Elodia, Hydrilla, Elodea, Potamogeton, Myriophyllum, Ranunculus, Utricularia, Ceratophyllum, Vallisnena, Chara etc.
- Floating stage: When the depth of water reaches about 4 to 8 feet, the submerged vegetation starts disappearing and then the floating plants make their appearance gradually in that area. Their broad leaves floating on the water surface check the penetration of light to deeper layer of water. Due to continuous interaction between plant communities and aquatic environment, the habitat becomes changed chemically as well as physically. The substratum rises up in vertical direction. Important floating plants that replace the submerged vegetation are Nelumbmm, Trapa, Pistia, Nymphaea, Wolffia, Lemna, Aponogeton and Limnanthemum etc.
- Reed swamp stages: The floating plants start disappearing gradually and their places are occupied by amphibious plants which can live successfully in aquatic as well as aerial environment. Important examples are Bothrioclova, Typha, Phragmites, Scripus, Sagittaria etc. The foliage leaves of such plants are exposed much above the surface of water and roots are generally found either in mud or submerged in water.

- Sedge Meadow stage: The filling process finally results in a marshy soil which may be too dry for the plants of pre-existing community. The members of cyperaceae and grammeae appeared. The species of Carex, Juncus, Themeda, Iris, Dichanthium, Eriophorum, Cymbopogon, Campanula, Mentha, Caltha, Gallium, Teucrium, polygonum, Eleocharis, Cicuta, etc.
- Woodland stage: In the beginning some shrubs and later medium sized trees form open vegetation or woodland. These plants produce more shade and absorb and transpire large quantity of water. The prominent plants of woodland community are species of Buteazon, Cornus, Populus, Almus, Acacia, Cassia, Terminalia, Salix, Cephalanthus, etc.
- Climax forest: After a very long time the hydrosere may lead to the development of climax vegetation, well adapted self-maintaining and self-reproducing. In the climax forest, all types of plants are present herbs, shrubs, mosses and shade loving plants represent their own communities. Trees are dominant and they have control over the entire vegetation. Bacteria, fungi, and other micro-organisms are more frequently found in the climax vegetation.

Succession of plant is followed by succession of animal: Fish-> Insect, Prawn, Snail-> Birds-> Land animals. Xerosere involves the ecological succession on bare rock surface. The rocky habitat shows many extreme xeric conditions like very high surface temperature, deficiency of water and absence of organic matter.

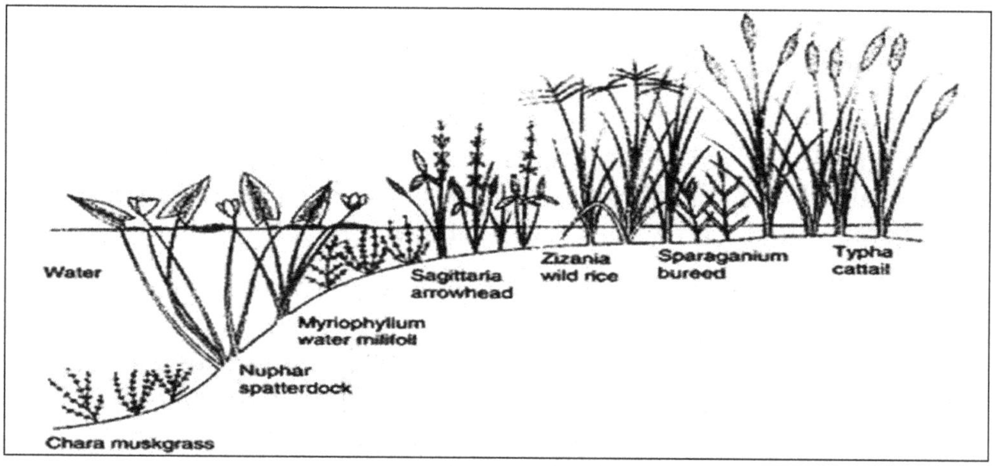

Community succession along a pod and along river banks.

The various stages of successions on bare area are:

- Crustose Lichen Stage: The pioneer colonisers on the bare area are crustose lichens which occur on the rock surface in the form of membranous crusts. Important crustose lichens are Rhizocarpon, Rhinodina, Graphis, Lacidea, Lacanora etc. Their spongy nature enables them absorb excess amount of water and minerals. These lichens migrate through their spores and soridia and their migration is facilitated by wind and water. The lichens secrete carbonic acid in excess.
- Foliose Lichen Stage: After accumulation of little soil and humus, the rock surface now becomes covered with xeric foliose and fructicose lichens, like Dermatocarpon, Parmellia, Umbilicaria, etc. These lichens have delicate leaf like thalli which cover the rocks. When the supply of light is cut off the crustose lichens begin to die. Foliose lichens absorb and accumulate water and minerals and check evaporation of surface water. They also secrete carbonic acid which further pulverizes or loosens the rocks into small particles.

- Moss Stage: The existing foliose lichens start disappearing in that area xerophytic mosses like Polytrichum, Torulla, Grimmia etc. grow in the crevices and the depressions of the rocks and become dominant. The xerophytic mosses develop rhizoids which penetrate deep into the rocky soil. The decaying older parts of mosses form a thick mat over the rock surface, which increases the water holding capacity of soil.
- Herbs Stage: The herbaceous vegetation, mainly consists of annual herbs (Eleusine, Aristida) and perennial herbs (Cymbopogon, Heteropogon), develops very quickly. Increased moisture content of the soil favours the growth of herbs. The roots of these plants penetrate down almost to the level of unpulverized rock. Decaying leaves stems, roots and other parts of the plants the form of humus and increase the water holding capacity of the soil.
- Shrub Stage: Xerophytic shrubs gradually occupy the area. Roots of shrubs also reach the surface of unpulverized rocks and corrode sufficient quantity of rock particles which make the soil more massive. Decaying leaves, twigs and roots of these shrubs also enrich the soil with humus. Some important shrubs are Rhus, Phytocarpus, Zizyphus, Caparis.
- Forest Stage: The xerophytic trees invade the area which has been occupied previously by shrubs. The first trees growing in such areas are dwarf and widely spaced. Then mesophytic trees grow densely and become dominant. In the shade of mesophytic trees some shade loving herbs and shrubs which are well adapted to humid atmosphere also appear and they form their own communities. The climax stage established which remains unchanged unless some major environmental changes disturb it.

Succession of plant is followed by succession of animal: Spider and grasshopper -> Ants, Termites, Earthworm, Snails, Centipede, Millipede, Birds, and squirrels.

References

- What-is-an-ecosystem: internetgeography.net, Retrieved 12, May 2020
- Ecosystem: biologydictionary.net, Retrieved 10, June 2020
- Abiotic-components, biology: vedantu.com, Retrieved 17, July 2020
- Biotic-factors: biologydictionary.net, Retrieved 22, March 2020
- Ecological-Niche-323908147: researchgate.net, Retrieved 14, January 2020

Chapter 3

Ecosystem Ecology

The living components of an ecosystem can be organized into producers and consumers. They form an intricate association with each other through food chains and food webs. Energy flows through the ecosystem through these chains in a process called the energy cycle. The topics elaborated in this chapter will help in gaining a better perspective of energy flow in ecosystems.

Ecosystem ecology is the combined study of the physical and biological components of ecosystems. It focuses on how matter and energy flow through both organisms and the abiotic components of the environment.

Ecosystem ecology examines large-scale ecological issues, ones that often are framed in terms not of species but rather of measures such as biomass, energy flow, and nutrient cycling. Questions include how much carbon is absorbed from the atmosphere by terrestrial plants and marine phytoplankton during photosynthesis and how much of that is consumed by herbivores, the herbivores' predators, and so on up the food chain. Carbon is the basis of life, so these questions may be framed in terms of energy. How much food one has to eat each day, for instance, can be measured in terms of its dry weight or its calorie content. The same applies to measures of production for all the plants in an ecosystem or for different trophic levels of an ecosystem. A basic question in ecosystem ecology is how much production there is and what the factors are that affect it. Not surprisingly, warm, wet places such as rainforests produce more than extremely cold or dry places, but other factors are important. Nutrients are essential and may be in limited supply. The availability of phosphorus and nitrogen often determines productivity—it is the reason these substances are added to lawns and crops—and their availability is particularly important in aquatic systems. On the other hand, nutrients can represent too much of a good thing. Human activity has modified global ecosystems in ways that are increasing atmospheric carbon dioxide, a carbon source but also a greenhouse gas, and causing excessive runoff of fertilizers into rivers and then into the ocean, where it kills the species that live there.

Energy Flow

Living organisms can use energy in two forms radiant and fixed energy. Radiant energy is in the form of electromagnetic waves, such as light. Fixed energy is potential chemical energy bound in various organic substances which can be broken down in order to release their energy content.

Organisms that can fix radiant energy utilizing inorganic substances to produce organic molecules are called autotrophs. Organisms that cannot obtain energy from abiotic source but depend on energy-rich organic molecules synthesized by autotrophs are called heterotrophs. Those which obtain energy from living organisms are called consumers and those which obtain energy from dead organisms are called decomposers.

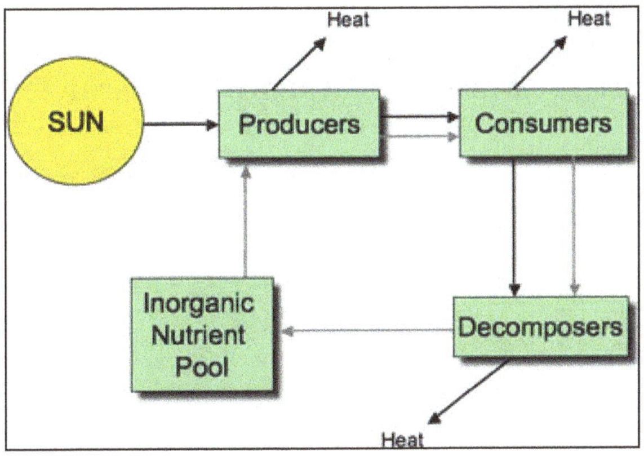

Flow of energy: Diffrent levels of ecosystem.

When the light energy falls on the green surfaces of plants, a part of it is transformed into chemical energy which is stored in various organic products in the plants. When the herbivores consume plants as food and convert chemical energy accumulated in plant products into kinetic energy, degradation of energy will occur through its conversion into heat. When herbivores are consumed by carnivores of the first order (secondary consumers) further degradation will occur. Similarly, when primary carnivores are consumed by top carnivores, again energy will be degraded.

Trophic Level

The producers and consumers in ecosystem can be arranged into several feeding groups, each known as trophic level (feeding level). In any ecosystem, producers represent the first trophic level, herbivores present the second trophic level, primary carnivores represent the third trophic level and top carnivores represent the last level.

Food Chain

In the ecosystem, green plants alone are able to trap in solar energy and convert it into chemical energy. The chemical energy is locked up in the various organic compounds, such as carbohydrates, fats and proteins, present in the green plants. Since virtually all other living organisms depend upon green plants for their energy, the efficiency of plants in any given area in capturing solar energy sets the upper limit to long-term energy flow and biological activity in the community.

The food manufactured by the green plants is utilized by themselves and also by herbivores. Animals feed repeatedly. Herbivores fall prey to some carnivorous animals. In this way one form of life supports the other form. Thus, food from one trophic level reaches to the other trophic level and in this way a chain is established. This is known as food chain.

A food chain may be defined as the transfer of energy and nutrients through a succession of organisms through repeated process of eating and being eaten. In food chain initial link is a green plant or producer which produces chemical energy available to consumers. For example, marsh grass is consumed by grasshopper, the grasshopper is consumed by a bird and that bird is consumed by hawk.

Thus, a Food Chain is Formed which can be Written as Follows

Marsh grass → grasshopper → bird → hawk

Food chain in any ecosystem runs directly in which green plants are eaten by herbivores, herbivores are eaten by carnivores and carnivores are eaten by top carnivores. Man forms the terrestrial links of many food chains.

Food Chains are of Three Types

1. Grazing food chain.
2. Parasitic food chain.
3. Saprophytic or detritus food chain.

1. Grazing Food Chain

The grazing food chain starts from green plants and from autotrophs it goes to herbivores (primary consumers) to primary carnivores (secondary consumers) and then to secondary carnivores (tertiary consumers) and so on. The gross production of a green plant in an ecosystem may meet three fates—it may be oxidized in respiration, it may be eaten by herbivorous animals and after the death and decay of producers it may be utilized by decomposers and converters and finally released into the environment. In herbivores the assimilated food can be stored as carbohydrates, proteins and fats, and transformed into much more complex organic molecules.

The energy for these transformations is supplied through respiration. As in autotrophs, the energy in herbivores also meets three routes respiration, decay of organic matter by microbes and consumption by the carnivores. Likewise, when the secondary carnivores or tertiary consumers eat primary carnivores, the total energy assimilated by primary carnivores or gross tertiary production follows the same course and its disposition into respiration, decay and further consumption by other carnivores is entirely similar to that of herbivores.

Thus, it is obvious that much of the energy flow in the grazing food chain can be described in terms of trophic levels as outlined below:

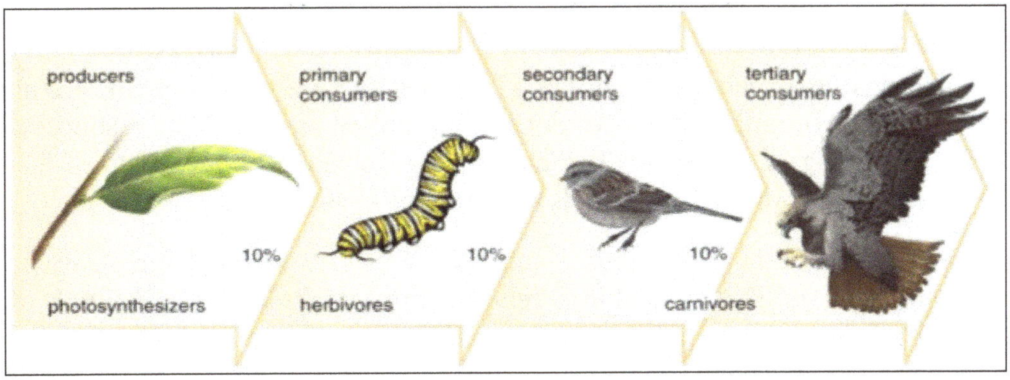

A schematic representation of grazing food chain showing input and losses of energy has been presented in figure.

2. Parasitic Food Chain

It goes from large organisms to smaller ones without outright killing as in the case of predator.

3. Detritus Food Chain

The dead organic remains including metabolic wastes and exudates derived from grazing food chain are generally termed detritus. The energy contained in detritus is not lost in ecosystem as a whole, rather it serves as a source of energy for a group of organisms called detritivores that are separate from the grazing food chain. The food chain so formed is called detritus food chain.

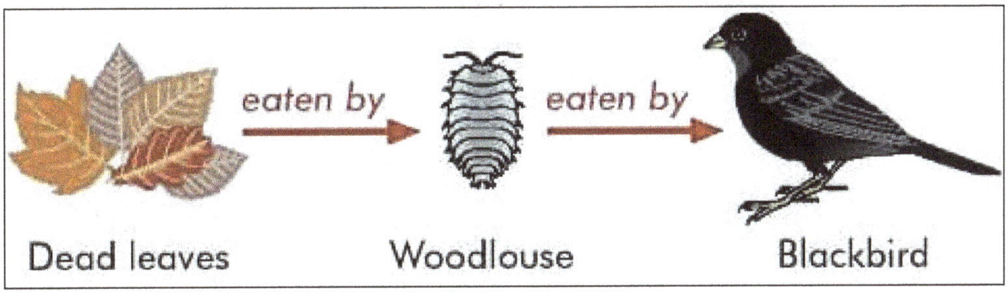

In some ecosystems more energy flows through the detritus food chain than through grazing food chain. In detritus food chain the energy flow remains as a continuous passage rather than as a stepwise flow between discrete entities. The organisms in the detritus food chain are many and include algae, fungi, bacteria, slime moulds, actinomycetes, protozoa, etc. Detritus organisms ingest pieces of partially decomposed organic matter, digest them partially and after extracting some of the chemical energy in the food to run their metabolism, excrete the remainder in the form of simpler organic molecules.

The waste from one organism can be immediately utilized by a second one which repeats the process. Gradually, the complex organic molecules present in the organic wastes or dead tissues are broken down to much simpler compounds, sometimes to carbon dioxide and water and all that are left are humus. In a normal environment the humus is quite stable and forms an essential part of the soil.

Food Web

Many food chains exist in an ecosystem, but as a matter of fact these food chains are not independent. In ecosystem, one organism does not depend wholly on another. The resources are shared specially at the beginning of the chain. The marsh plants are eaten by variety of insects, birds, mammals and fishes and some of the animals are eaten by several predators.

Similarly, in the food chain grass → mouse → snakes → owls, sometimes mice are not eaten by snakes but directly by owls. This type of interrelationship interlinks the individuals of the whole community. In this way, food chains become interlinked. A complex of interrelated food chains makes up a food web. Food web maintains the stability of the ecosystem. The greater the number of alternative pathways the more stable is the community of living things. Figure below illustrates a food web in ecosystem.

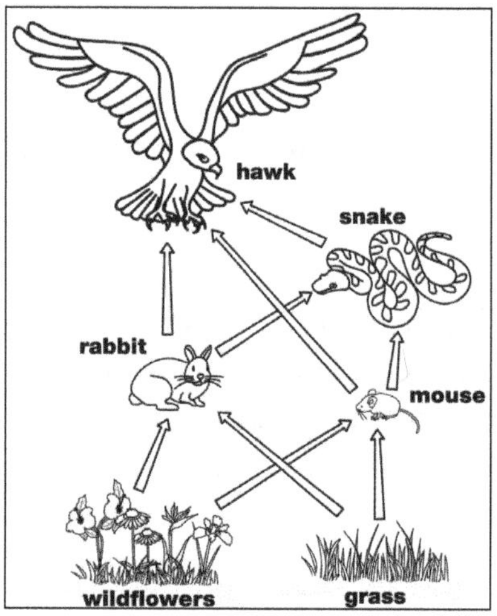

Food web in an ecosystem.

Ecological Pyramid

The trophic structure of an ecosystem can be indicated by means of ecological pyramid. At each step in the food chain a considerable fraction of the potential energy is lost as heat. As a result, organisms in each trophic level pass on lesser energy to the next trophic level than they actually receive. This limits the number of steps in any food chain to 4 or 5. Longer the food chain the lesser energy is available for final members. Because of this tapering off of available energy in the food chain a pyramid is formed that is known as ecological pyramid. The higher the steps in the ecological pyramid the lower will be the number of individuals and the larger their size.

The idea of ecological pyramids was advanced by C.E. Elton (1927). There are different types of ecological pyramids. In each ecological pyramid, producer level forms the base and successive levels make up the apex. Three types of pyramidal relations may be found among the organisms at different levels in the ecosystem.

These are as Follows

1. Pyramid of numbers.

2. Pyramid of biomass (biomass is the weight of living organisms), and
3. Pyramid of energy.

1. Pyramid of Numbers

It depicts the numbers of individuals in producers and in different orders of consumers in an ecosystem. The base of pyramid is represented by producers which are the most abundant. In the successive levels of consumers, the number of organisms goes on decreasing rapidly until there are a few carnivores.

The pyramid of numbers of an ecosystem indicates that the producers are ingested in large numbers by smaller numbers of primary consumers. These primary consumers are eaten by relatively smaller number of secondary consumers and these secondary consumers, in turn, are consumed by only a few tertiary consumers.

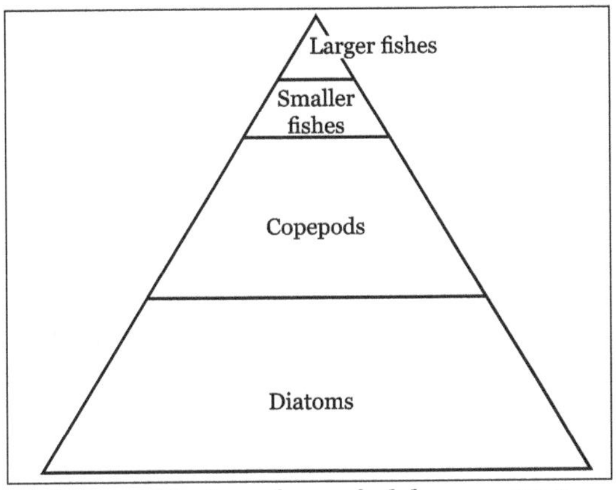

A Pyramid of numbers of a lake ecosystem.

This type of pyramid is best presented by taking an example of Lake Ecosystem. In this type of pyramid the base trophic level is occupied by producer elements—algae, diatoms and other hydrophytes which are most abundant. At the second trophic level come the herbivores or zooplanktons which are lesser in number than producers.

The third trophic level is occupied by carnivores which are still smaller in number than the herbivores and the top is occupied by a few top carnivores. Thus, in the ecological pyramid of numbers there is a relative reduction in number of organisms and an increase in the size of body from base to apex of the pyramid. In parasitic food chain starting from tree, the pyramid of numbers will be inverted.

(A) Up-right Pyramids of numbers in a grassland and cultivated field
(B) Pyramids of numbers (inverted) of diseased tree (Parasitic ecosystem).

2. Pyramid of Biomass of Organisms

The living weights or biomass of the members of the food chain present at any one time form the pyramid of biomass of organisms. This indicates, by weight or other means of measuring materials, the total bulk of organisms or fixed energy present at one time. Pyramid of biomass indicates the decrease of biomass in each tropic level from base to apex, e.g., total biomass of producers is more than the total biomass of the herbivores.

Likewise, the total biomass of secondary consumers will be lesser than that of herbivores and so on. Since some energy and material are lost in each successive link, the total mass supported at each level is limited by the rate at which the energy is being stored below. This usually gives sloping pyramid for most of the communities in terrestrial and shallow water ecosystems. The pyramid of biomass in a pond ecosystem will be inverted as shown in figure below.

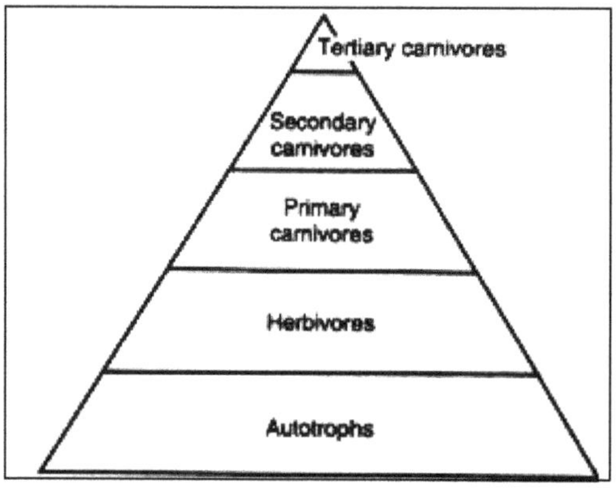

A pyramid of biomass.

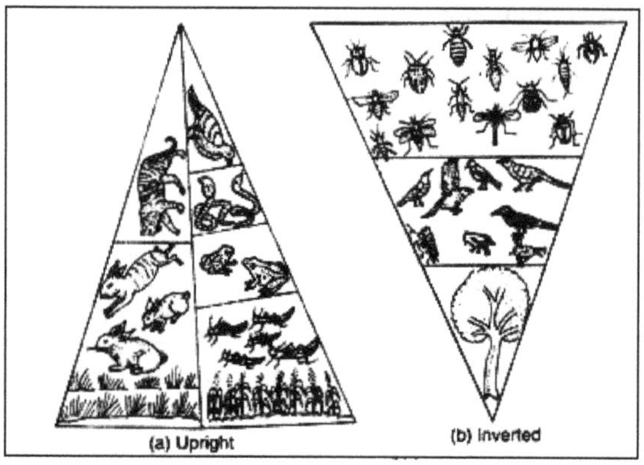

A pyramid of biomass.

3. Pyramid of Energy

This depicts not only the amount of total energy utilized by the organisms at each trophic level of food chain but more important, the actual role of various organisms in transfer of energy. At the producer level the total energy will be much greater than the energy at the successive higher trophic level.

Some producer organisms may have small biomass but the total energy they assimilate and pass on to consumers may be greater than that of organisms with much larger biomass. Higher trophic levels are more efficient in energy utilization but much heat is lost in energy transfer. Energy loss by respiration also progressively increases from lower to higher trophic states.

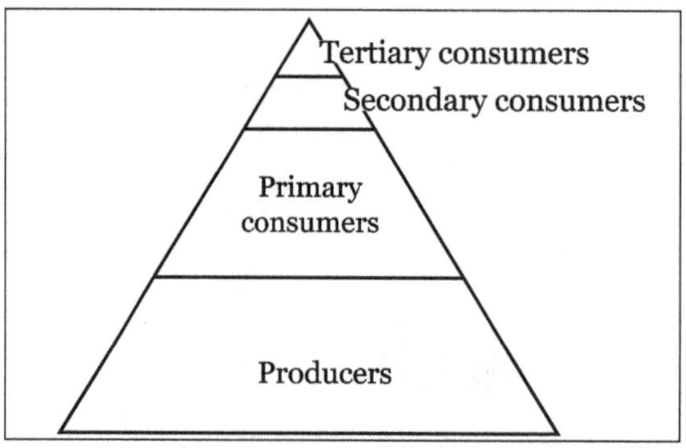

Figure: A pyramid of energy.

In the energy flow process, two things become obvious. Firstly there is one way along which energy moves i.e. unidirectional flow of energy. Energy comes in the ecosystem from outside source i.e. sun. The energy captured by autotrophs does not go back to the sun, the energy that passes from autotrophs to herbivores does not revert back and as it moves progressively through the various trophic levels, it is no longer available to the previous levels.

Thus due to unidirectional flow of energy, the system would collapse if the supply from primary source, the sun is cut off. Secondly, there occurs a progressive decrease in energy level at each trophic level which is accounted largely by the energy dissipated as heat in metabolic activities.

Productivity

The relationship between the amount of energy accumulated and the amount of energy utilized within one trophic level of food chain has an important bearing on how much energy from one trophic level passes on to the next trophic level in the food chain. The ratio of output of energy to input of energy is referred to as ecological efficiency.

Different kinds of Efficiencies can be Measured by the Following Parameters

(i) Ingestion which indicates the quantity of food or energy taken by trophic level. This is also called exploitation efficiency.

(ii) Assimilation indicates the amount of food absorbed and fixed into energy rich organic substances which are stored or combined with other molecules to build complex molecules such as proteins, fats etc.

(iii) Respiration which indicates the energy lost in metabolism.

Primary Productivity

The fraction of fixed energy a trophic level passes on to the next trophic level is called production. Green plants fix solar energy and accumulate it in organic forms as chemical energy. Since it is the first and basic form of energy storage, the rate at which the energy accumulates in the green plants or producers is known as primary productivity.

Primary productivity is the rate at which energy is bound or organic material is created by photosynthesis per unit area of earth's surface per unit time. It is most often expressed as energy in calories / cm^2 / yr or dry organic matter in g / m^2 / yr (g/m^2 x 8.92 = lb / acre). The amount of organic matter present at a given time per unit area is called standing crop or biomass and as such productivity, which is a rate, is quite different from biomass or standing crop.

The standing crop is usually expressed as dry weight in g/m^2 or kg/m^2 or t/ha (metric tons) or 10^6g/hectare. Primary productivity is the result of photosynthesis by green plants including algae of different colours. Bacterial photosynthesis or chemosynthesis, although of small significance may also contribute to primary productivity. The total solar energy trapped in the food material by photosynthesis is referred to as gross primary productivity (G.P.P.).

A good fraction of gross primary production is utilized in respiration of green plants. The amount of energy bound in organic matter per unit area and time that is left after respiration in plants is net primary production (N.P.P.) or plant growth. Only the net primary productivity is available for harvest by man and other animals. Net productivity of energy = gross productivity—energy lost in respiration.

Secondary Productivity

The rates at which the heterotrophic organisms resynthesize the energy-yielding substances is termed as secondary productivity. Secondary productivities are the productivities of animals and saprobes in communities. The amount of energy stored in the tissues of consumers or heterotrophs is termed as net secondary production and the total plant material ingested by herbivores is grass secondary production. Total plant material ingested by herbivores minus the materials lost as faeces is equal to Ingested Secondary Production.

Environmental Factors Affecting the Production Processes in an Ecosystem are as Follows

1. Solar radiation and Temperature.
2. Moisture. Leaf water potential, soil moisture and precipitation fluctuation and transpiration.
3. Mineral nutrition. Uptake of minerals from the soil, rhizosphere effects, fire effects, salinity, heavy metals, nitrogen metabolism.
4. Biotic activities. Grazing, above ground herbivores, below ground herbivores, predators and parasites, diseases of primary producers.
5. Impact of human population. Pollutions of different sorts, ionizing radiations like atomic explosions, etc.

There are three Fundamental Concepts of Productivity

1. Standing crop.
2. Materials removed.
3. Production rate.

1. Standing Crop

It is the abundance of the organisms existing in the area at any one time. It may be expressed in terms of number of individuals, as biomass of organisms, as energy content or in some other suitable terms. Measurement of standing crop reveals the concentration of individuals in the various populations of ecosystem.

2. The Materials Removed

The second concept of productivity is the materials removed from the area per unit time. It includes the yield to man, organisms removed from the ecosystem by migration, and the material withdrawn as organic deposit.

3. The Production Rate

The third concept of productivity is the production rate. It is the rate at which the growth processes are going forward within the area. The amount of material formed by each link in the food chain per unit of time per unit area or volume is the production rate.

All the three major groups of organisms—producers, consumers and reducers are the functional kingdoms of natural communities. The three represent major directions of evolution and are characterised by different modes of nutrition. Plants feed primarily by photosynthesis, animals feed primarily by ingesting food that is digested and absorbed in the alimentary canal and the saprobes feed by absorption and have need for an extensive surface of absorption. The principal kinds of organisms among saprobes are the unicellular bacteria, yeasts, chytrids or lower fungi and higher fungi with mycelial bodies.

In terrestrial communities as much as 90% of net primary production remains un-harvested and are utilized as dead tissue by saprobes and soil animals. The saprobes have a larger and more essential role than animals in degrading dead organic matter to inorganic forms and in such ecosystems, secondary production by reducers (decomposers) should exceed that by consumers, though the former is even more difficult to measure than the latter.

Biomass of decomposers with their microscopic cells and filaments embedded in food sources is also difficult to measure and that is small in relation to their productivity and significance for the ecosystem. Small masses of reducers degrade and transform larger masses of organic matter to inorganic remnants. In so doing decomposers disperse back to the environment the energy of photosynthesis accumulated in the organic compounds that are decomposed.

Thus they have a major role in the energy flow of ecosystems. A community or ecosystem, like an organism, is an open energy system. The continuous intake of energy in photosynthesis replaces the energy dissipated to environment by respiration and biological activity and the system does not run-down through the loss of free energy to maximum entropy.

If the amount of energy entrapped is greater than the energy dissipated, the pool of biologically useful energy of organic bonds increases. This results in increase of community biomass and consequently the community grows; such is the case in succession. If energy intake is lesser than energy dissipation, the community biomass will decrease and it must, in some sense, retrogress. If energy intake and loss are in balance, the pool of organic energy is in steady state; such is the case in climax communities.

Three Aspects of this Steady State may be Recognized

(i) The steady state of population of climax communities in which equal birth and death rates in population keep the number of individuals relatively constant.

(ii) The steady state of energy flow.

(iii) The steady state of the matter of community, where addition of material by photosynthesis and organic synthesis is balanced by loss of material through respiration and decomposition.

Methods of Measuring Primary Production

There are several parameters for measuring primary production and the methods of measuring primary production are based on those parameters.

The Methods are Discussed here as Under

1. Harvest Method

It involves removal of vegetation periodically and weighing the material. For measuring above ground production, the above ground plant parts are clipped at ground level, dried to constant weight at 80°C and weighed. The dry weight in g/m² /year gives the ground production. Below ground production is estimated by using frequent core sampling technique of Dahlman and Kucera (1965). It is expressed in terms of weight in gm per unit area per year. In terms of energy one gm dry weight of plant material contains 4 to 5 kcal.

The Limitations of Harvest Method are as Follows

(i) The amount of plant material consumed by herbivores and the food oxidized during respiration process of the plants is not accounted.

(ii) Root biomass is neglected.

(iii) Photosynthetic trans located to underground parts of plants are not known.

In spite of these limitations the method is used all over for measuring net assimilation rate (NAR) and relative growth rate (RGR).

2. Carbon Dioxide Assimilation Method

Utilization of CO_2 in photosynthesis or its liberation during respiration is measured by infrared gas analysis or by passing the gas through Baryta water $Ba(OH)_2$ and titrating the same. The CO_2 removed from incoming gas chamber is taken to be synthesized into organic matter by the green plants. Performing the experiment in light and dark chambers the net and gross production can be measured.

In the lighted chamber photosynthesis and respiration take place simultaneously and the CO_2 coming out from the chamber is the unused gas of the atmosphere plus gas from the respiration of plant parts. In the dark chamber all CO_2 is due to respiration.

Net production = Gross production—Respiration

3. Oxygen Production Method

In the aquatic vegetation CO_2 gas analysis method is not used but oxygen evolution method is generally used. The light and dark bottle technique is employed for measuring primary production of aquatic plant. In this method two bottles, one transparent and the other opaque are filled with water at a given depth of lake, closed, maintained at that depth for some time and then brought to laboratory for determination of oxygen content in the water. The decrease of oxygen in dark bottle is due to respiratory activity while increase of O_2 in light bottle is due to photosynthesis. The total increase of O_2 in light bottle plus the amount of O_2 decreased in dark bottle express gross productivity (O_2 value multiplied by 0.375 gives an equivalent of carbon assimilation). Recently, oxygen electrodes have been used for estimating oxygen content in water.

4. Chlorophyll Method

Gesner (1949) pointed out that the amount of chlorophyll/m^2 is almost limited to a narrow range of 0.1 to 3.0 gm regardless of the age of individuals or the species present therein. There is direct correlation between the amount of chlorophyll and dry matter production in different types of communities with varying light conditions.

The relation of total amount of chlorophyll to the photosynthetic rate is referred to as assimilation ratio or rate of production/gm chlorophyll. Total chlorophyll per unit area is greater in land plants as compared to that in aquatic plants. In marine ecosystem the rate of carbon assimilation is 3.7 g/ hr/g of chlorophyll. The relationship between area based chlorophyll and dry matter production in terrestrial ecosystems has been worked out by Japanese ecologists Argua and Monsi (1963).

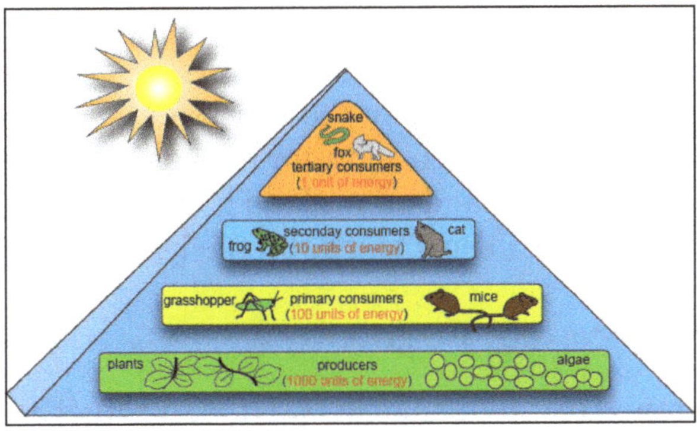

Ecosystem Engineer

Organisms impact the environment in which they live in in several ways. Some organisms are destructive to the ecosystem while others contribute positively to the development of the ecosystem. Organisms that create, modify, destroy or maintain a habitat in which they live or frequent are known as ecosystem engineers. These organisms can have a great impact on the species-richness and heterogeneity of the landscape of an area. Ecosystem Engineers maintain the health and stability of the environment they live in. However, since all organisms contribute to the modification of their environment in one way or the other, ecosystem engineer is only used to describe keystone species which plays a critical role in maintaining the ecological community and affects other organisms in the ecosystem.

Types of Ecosystem Engineers

Ecosystem engineers are divided into two broad categories: allogenic and autogenic engineers. Allogenic engineers modify the environment by mechanically transforming material, both living and non-living, from one form to another or various forms. Beavers, which are the original model for ecosystem engineers, alter their ecosystem extensively through the process of clear-cutting and damming. The addition of the dam changes the distribution and the abundance of the organisms in the area. By creating a shelter from leaves, a caterpillar also creates a shelter for another organism which may interact with it simultaneously or as a result of building the shelter. Birds that create holes in trees and wood to nest in also create homes for other organisms once they are through with them. Autogenic engineers, on the other hand, modify the environment on which they are in by modifying themselves. Trees are the best examples of autogenic engineers since as they grow, their trunks, leaves, and branches are used by other organisms as habitats including birds, insects, snakes, and other organisms. Trees may also form a forest which is a suitable habitat for other organisms.

Importance of Ecosystem Engineers

Ecosystem engineers have greater influence and impact on other organisms living in the same environment with them, especially by providing resources to the organisms. Some of the ecosystem engineers have contributed to species richness at the landscape level. The dams built by beavers have ecological effects on other species since they create habitat and control a number of abiotic resources that other animals can use and can also support species not found anywhere. By conserving ecosystem engineers, protection is extended to the overall diversity of a landscape. The biodiversity of an area can also be affected by the ecosystem engineer's ability to increase the complexity of the ecosystem processes which allow for further species greatness. Beavers have the ability to modify a riparian land and expand on wetland habitat leading to an increase in diversity of habitat creating room for more organisms to inhabit the landscape.

Can Humans also be Ecosystem Engineers?

Humans are the most obvious ecosystem engineers. Human activities have contributed to niche construction. The activities such as agriculture, mining, logging and other activities have significantly changed how humans interact with the environment. Some of the human intervention activities such as environmental preservation and upgrading enables an area to be restored to the previous state. Humans also help in the management of evasive species that would otherwise be dangerous to the environment and organisms inhabiting such area.

Productivity

The productivity of an ecosystem refers to the rate of production i.e. the amount of organic matter accumulated in any unit time. Productivity is of following types:

1. Primary Productivity. It is associated with the producers which are autotrophic, most of which are photosynthetic, and to a much lesser extent the chemosynthetic microorganisms. These are the green plants, higher macrophytes as well as lower forms, the phytoplanktons and some photosynthetic bacteria. Primary productivity is defined as "the rate at which radiant energy is stored by photosynthetic and chemosynthetic activities of the producers." Primary productivity is further distinguished as:

(a) Gross Primary Productivity. It is the total rate of photosynthesis including the organic matter used up in respiration during the measurement period. This is also sometimes referred to as Total (Gross) Photosynthesis or Total Assimilation. It depends on the chlorophyll content. It is estimated in terms of either chlorophyll content as, Chl/g dry weight/unit area, or photosynthetic number i.e. amount of CO_2 fixed/g Chl/hour.

(b) Net primary Productivity. It is the rate of storage of organic matter in plant tissues in excess of the respiratory utilization by plants during the measurement period. This is thus the rate of increase of biomass and is also known as Apparent Photosynthesis or Net assimilation. Thus net primary productivity refers to balance between gross photosynthesis and respiration and other plant losses as death etc.

2. Secondary Productivity. It refers to the consumers or heterotrophs. These are the rates of energy storage at consumer level. Since consumers only utilize food material (already produced) in their respiration, simply converting the food matter to different tissues by an overall process, secondary productivity is not divided into 'gross' and 'net' amounts. Thus, some ecologists prefer to use the term assimilation rather than production at this level. Secondary productivity actually remains mobile (i.e. keeps moving from one organism to another) and does not live in situ like the primary productivity.

3. Net Productivity. It refers to the rate of storage of organic matter not used by the heterotrophs (consumers) i.e. equivalent to net primary production minus consumption by the heterotrophs during the unit period, as a season or year etc. It is thus the rate of increase of biomass of the primary producers which has been left over by the consumers. Net productivity is generally expressed as production of C g/m^2/day which may then be consolidated on month, season or year basis.

Nutrient Cycle

Nutrient cycling is one of the most important processes that occur in an ecosystem. The nutrient cycle describes the use, movement, and recycling of nutrients in the environment. Valuable elements such as carbon, oxygen, hydrogen, phosphorus, and nitrogen are essential to life and must be recycled in order for organisms to exist. Nutrient cycles are inclusive of both living and non-living components and involve biological, geological, and chemical processes. For this reason, these nutrient circuits are known as biogeochemical cycles.

Biogeochemical Cycles

Biogeochemical cycles can be categorized into two main types: global cycles and local cycles. Elements such as carbon, nitrogen, oxygen, and hydrogen are recycled through abiotic environments including the atmosphere, water, and soil. Since the atmosphere is the main abiotic environment from which these elements are harvested, their cycles are of a global nature. These elements may travel over large distances before they are taken up by biological organisms. The soil is the main abiotic environment for the recycling of elements such as phosphorus, calcium, and potassium. As such, their movement is typically over a local region.

Carbon Cycle

Carbon is essential to all life as it is the main constituent of living organisms. It serves as the backbone component for all organic polymers, including carbohydrates, proteins, and lipids. Carbon compounds, such as carbon dioxide (CO_2) and methane (CH_4), circulate in the atmosphere and influence global climates. Carbon is circulated between living and nonliving components of the ecosystem primarily through the processes of photosynthesis and respiration. Plants and other photosynthetic organisms obtain CO_2 from their environment and use it to build biological materials. Plants, animals, and decomposers (bacteria and fungi) return CO_2 to the atmosphere through respiration. The movement of carbon through biotic components of the environment is known as the fast carbon cycle. It takes considerably less time for carbon to move through the biotic elements of the cycle than it takes for it to move through the abiotic elements. It can take as long as 200 million years for carbon to move through abiotic elements such as rocks, soil, and oceans. Thus, this circulation of carbon is known as the slow carbon cycle.

Carbon Cycles Through the Environment as Follows:

- CO_2 is removed from the atmosphere by photosynthetic organisms (plants, cyanobacteria, etc.) and used to generate organic molecules and build biological mass.
- Animals consume the photosynthetic organisms and acquire the carbon stored within the producers.
- CO_2 is returned to the atmosphere via respiration in all living organisms.
- Decomposers break down dead and decaying organic matter and release CO_2.
- Some CO_2 is returned to the atmosphere via the burning of organic matter (forest fires).
- CO_2 trapped in rock or fossil fuels can be returned to the atmosphere via erosion, volcanic eruptions, or fossil fuel combustion.

Nitrogen Cycle

Similar to carbon, nitrogen is a necessary component of biological molecules. Some of these molecules include amino acids and nucleic acids. Although nitrogen (N_2) is abundant in the atmosphere, most living organisms can not use nitrogen in this form to synthesize organic compounds. Atmospheric nitrogen must first be fixed, or converted to ammonia (NH_3) by certain bacteria.

Nitrogen cycles through the environment as follows:

- Atmospheric nitrogen (N_2) is converted to ammonia (NH_3) by nitrogen-fixing bacteria in aquatic and soil environments. These organisms use nitrogen to synthesise the biological molecules they need to survive.
- NH_3 is subsequently converted to nitrite and nitrate by bacteria known as nitrifying bacteria.
- Plants obtain nitrogen from the soil by absorbing ammonium (NH_4-) and nitrate through their roots. Nitrate and ammonium are used to produce organic compounds.

- Nitrogen in its organic form is obtained by animals when they consume plants or animals.
- Decomposers return NH3 to the soil by decomposing solid waste and dead or decaying matter.
- Nitrifying bacteria convert NH3 to nitrite and nitrate.
- Denitrifying bacteria convert nitrite and nitrate to N2, releasing N2 back into the atmosphere.

Other Chemical Cycles

Oxygen and phosphorus are elements that are also essential to biological organisms. The vast majority of atmospheric oxygen (O2) is derived from photosynthesis. Plants and other photosynthetic organisms use CO2, water, and light energy to produce glucose and O2. Glucose is used to synthesize organic molecules, while O2 is released into the atmosphere. Oxygen is removed from the atmosphere through decomposition processes and respiration in living organisms.

Phosphorus is a component of biological molecules such as RNA, DNA, phospholipids, and adenosine triphosphate (ATP). ATP is a high energy molecule produced by the processes of cellular respiration and fermentation. In the phosphorus cycle, phosphorus is circulated mainly through soil, rocks, water, and living organisms. Phosphorus is found organically in the form of the phosphate ion (PO43-). Phosphorus is added to soil and water by runoff resulting from the weathering of rocks that contain phosphates. PO43- is absorbed from the soil by plants and obtained by consumers through the consumption of plants and other animals. Phosphates are added back to the soil through decomposition. Phosphates may also become trapped in sediments in aquatic environments. These phosphate containing sediments form new rocks over time.

Terrestrial Production

An oak tree; a typical modern, terrestrial autotroph.

On the land, almost all primary production is now performed by vascular plants, with a small fraction coming from algae and non-vascular plants such as mosses and liverworts. Before the evolution of vascular plants, non-vascular plants likely played a more significant role. Primary production on land is a function of many factors, but principally local hydrology and temperature (the latter covaries to an extent with light, specifically photosynthetically active radiation (PAR), the source of energy for photosynthesis). While plants cover much of the Earth's surface, they are strongly curtailed wherever temperatures are too extreme or where necessary plant resources (principally water and PAR) are limiting, such as deserts or polar regions.

Water is "consumed" in plants by the processes of photosynthesis and transpiration. The latter process (which is responsible for about 90% of water use) is driven by the evaporation of water from the leaves of plants. Transpiration allows plants to transport water and mineral nutrients from the soil to growth regions, and also cools the plant. Diffusion of water vapour out of a leaf, the force that drives transpiration, is regulated by structures known as stomata. These structure also regulate the diffusion of carbon dioxide from the atmosphere into the leaf, such that decreasing water loss (by partially closing stomata) also decreases carbon dioxide gain. Certain plants use alternative forms of photosynthesis, called Crassulacean acid metabolism (CAM) and C4. These employ physiological and anatomical adaptations to increase water-use efficiency and allow increased primary production to take place under conditions that would normally limit carbon fixation by C3 plants (the majority of plant species).

Oceanic Production

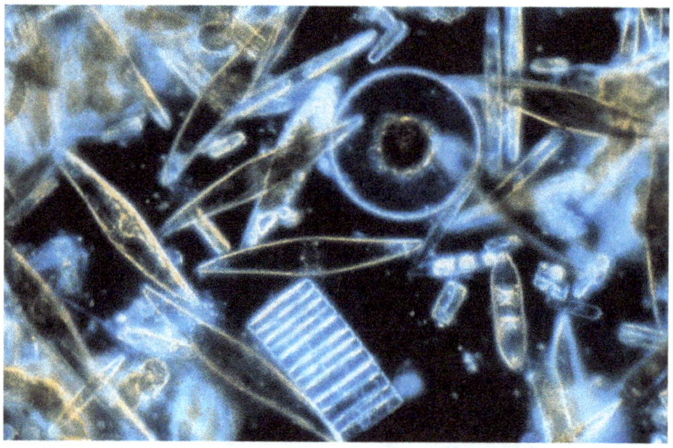

Marine diatoms; an example of planktonic microalgae.

In a reversal of the pattern on land, in the oceans, almost all photosynthesis is performed by algae, with a small fraction contributed by vascular plants and other groups. Algae encompass a diverse range of organisms, ranging from single floating cells to attached seaweeds. They include photoautotrophs from a variety of groups. Eubacteria are important photosynthetizers in both oceanic and terrestrial ecosystems, and while some archaea are phototrophic, none are known to utilise oxygen-evolving photosynthesis. A number of eukaryotes are significant contributors to primary production in the ocean, including green algae, brown algae and red algae, and a diverse group of unicellular groups. Vascular plants are also represented in the ocean by groups such as the seagrasses.

Unlike terrestrial ecosystems, the majority of primary production in the ocean is performed by free-living microscopic organisms called phytoplankton. Larger autotrophs, such as the seagrasses and macroalgae (seaweeds) are generally confined to the littoral zone and adjacent shallow waters, where they can attach to the underlying substrate but still be within the photic zone. There are exceptions, such as Sargassum, but the vast majority of free-floating production takes place within microscopic organisms.

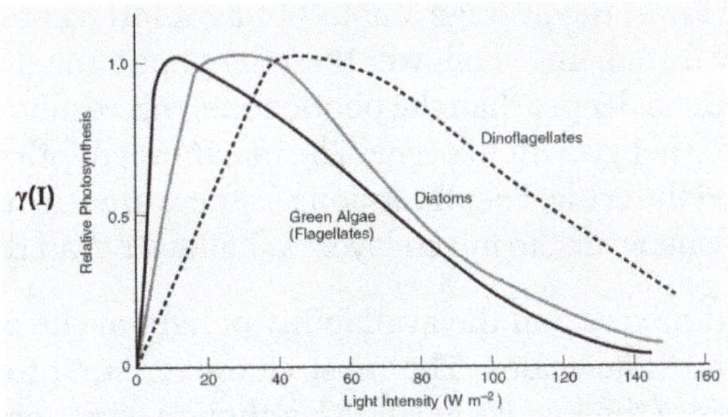

Differences in relative photosynthesis between plankton species under different irradiance.

The factors limiting primary production in the ocean are also very different from those on land. The availability of water, obviously, is not an issue (though its salinity can be). Similarly, temperature, while affecting metabolic rates, ranges less widely in the ocean than on land because the heat capacity of seawater buffers temperature changes, and the formation of sea ice insulates it at lower temperatures. However, the availability of light, the source of energy for photosynthesis, and mineral nutrients, the building blocks for new growth, play crucial roles in regulating primary production in the ocean. Available Earth System Models suggest that ongoing ocean bio-geochemical changes could trigger reductions in ocean NPP between 3% and 10% of current values depending on the emissions scenario.

Light

A kelp forest; an example of attached macroalgae.

The sunlit zone of the ocean is called the photic zone (or euphotic zone). This is a relatively thin layer (10–100 m) near the ocean's surface where there is sufficient light for photosynthesis to occur. For practical purposes, the thickness of the photic zone is typically defined by the depth at which light reaches 1% of its surface value. Light is attenuated down the water column by its absorption or scattering by the water itself, and by dissolved or particulate material within it (including phytoplankton).

Net photosynthesis in the water column is determined by the interaction between the photic zone and the mixed layer. Turbulent mixing by wind energy at the ocean's surface homogenises the water column vertically until the turbulence dissipates (creating the aforementioned mixed layer). The deeper the mixed layer, the lower the average amount of light intercepted by phytoplankton within it. The mixed layer can vary from being shallower than the photic zone, to being much deeper than the photic zone. When it is much deeper than the photic zone, this results in phytoplankton spending too much time in the dark for net growth to occur. The maximum depth of the mixed layer in which net growth can occur is called the critical depth. As long as there are adequate nutrients available, net primary production occurs whenever the mixed layer is shallower than the critical depth.

Both the magnitude of wind mixing and the availability of light at the ocean's surface are affected across a range of space- and time-scales. The most characteristic of these is the seasonal cycle (caused by the consequences of the Earth's axial tilt), although wind magnitudes additionally have strong spatial components. Consequently, primary production in temperate regions such as the North Atlantic is highly seasonal, varying with both incident light at the water's surface (reduced in winter) and the degree of mixing (increased in winter). In tropical regions, such as the gyres in the middle of the major basins, light may only vary slightly across the year, and mixing may only occur episodically, such as during large storms or hurricanes.

Nutrients

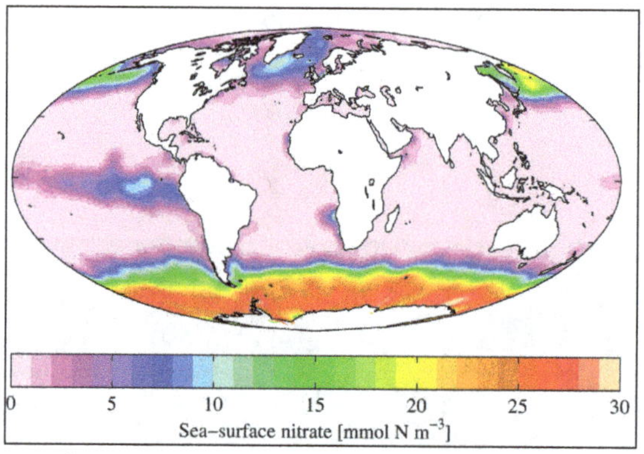

Annual mean sea surface nitrate for the World Ocean.

Mixing also plays an important role in the limitation of primary production by nutrients. Inorganic nutrients, such as nitrate, phosphate and silicic acid are necessary for phytoplankton to synthesise their cells and cellular machinery. Because of gravitational sinking of particulate material (such as plankton, dead or fecal material), nutrients are constantly lost from the photic zone, and are only replenished by mixing or upwelling of deeper water. This is exacerbated where summertime solar heating and reduced winds increases vertical stratification and leads to a strong thermocline, since this makes it more difficult for wind mixing to entrain deeper water. Consequently, between mixing events, primary production (and the resulting processes that leads to sinking particulate material) constantly acts to consume nutrients in the mixed layer, and in many regions this leads to nutrient exhaustion and decreased mixed layer production in the summer (even in the presence of abundant light). However, as long as the photic zone is deep enough, primary production may continue below the mixed layer where light-limited growth rates mean that nutrients are often more abundant.

Iron

Another factor relatively recently discovered to play a significant role in oceanic primary production is the micronutrient iron. This is used as a cofactor in enzymes involved in processes such as nitrate reduction and nitrogen fixation. A major source of iron to the oceans is dust from the Earth's deserts, picked up and delivered by the wind as aeolian dust.

In regions of the ocean that are distant from deserts or that are not reached by dust-carrying winds (for example, the Southern and North Pacific oceans), the lack of iron can severely limit the amount of primary production that can occur. These areas are sometimes known as HNLC (High-Nutrient, Low-Chlorophyll) regions, because the scarcity of iron both limits phytoplankton growth and leaves a surplus of other nutrients. Some scientists have suggested introducing iron to these areas as a means of increasing primary productivity and sequestering carbon dioxide from the atmosphere.

Measurement

The methods for measurement of primary production vary depending on whether gross vs net production is the desired measure, and whether terrestrial or aquatic systems are the focus. Gross production is almost always harder to measure than net, because of respiration, which is a continuous and ongoing process that consumes some of the products of primary production (i.e. sugars) before they can be accurately measured. Also, terrestrial ecosystems are generally more difficult because a substantial proportion of total productivity is shunted to below-ground organs and tissues, where it is logistically difficult to measure. Shallow water aquatic systems can also face this problem.

Scale also greatly affects measurement techniques. The rate of carbon assimilation in plant tissues, organs, whole plants, or plankton samples can be quantified by biochemically based techniques, but these techniques are decidedly inappropriate for large scale terrestrial field situations. There, net primary production is almost always the desired variable, and estimation techniques involve various methods of estimating dry-weight biomass changes over time. Biomass estimates are often converted to an energy measure, such as kilocalories, by an empirically determined conversion factor.

Terrestrial

In terrestrial ecosystems, researchers generally measure net primary production (NPP). Although its definition is straightforward, field measurements used to estimate productivity vary according to investigator and biome. Field estimates rarely account for below ground productivity, herbivory, turnover, litterfall, volatile organic compounds, root exudates, and allocation to symbiotic microorganisms. Biomass based NPP estimates result in underestimation of NPP due to incomplete accounting of these components. However, many field measurements correlate well to NPP. There are a number of comprehensive reviews of the field methods used to estimate NPP. Estimates of ecosystem respiration, the total carbon dioxide produced by the ecosystem, can also be made with gas flux measurements.

The major unaccounted pool is belowground productivity, especially production and turnover of roots. Belowground components of NPP are difficult to measure. BNPP (below-ground NPP) is often estimated based on a ratio of ANPP:BNPP (above-ground NPP:below-ground NPP) rather than direct measurements.

Gross primary production can be estimated from measurements of net ecosystem exchange (NEE) of carbon dioxide made by the eddy covariance technique. During night, this technique measures all components of ecosystem respiration. This respiration is scaled to day-time values and further subtracted from NEE.

Grasslands

The Konza tallgrass prairie in the Flint Hills of northeastern Kansas.

Most frequently, peak standing biomass is assumed to measure NPP. In systems with persistent standing litter, live biomass is commonly reported. Measures of peak biomass are more reliable if the system is predominantly annuals. However, perennial measurements could be reliable if there were a synchronous phenology driven by a strong seasonal climate. These methods may underestimate ANPP in grasslands by as much as 2 (temperate) to 4 (tropical) fold. Repeated measures of standing live and dead biomass provide more accurate estimates of all grasslands, particularly those with large turnover, rapid decomposition, and interspecific variation in timing of peak biomass. Wetland productivity (marshes and fens) is similarly measured. In Europe, annual mowing makes the annual biomass increment of wetlands evident.

Forests

Methods used to measure forest productivity are more diverse than those of grasslands. Biomass increment based on stand specific allometry plus litterfall is considered a suitable although incomplete accounting of above-ground net primary production (ANPP). Field measurements used as a proxy for ANPP include annual litterfall, diameter or basal area increment (DBH or BAI), and volume increment.

Aquatic

In aquatic systems, primary production is typically measured using one of six main techniques:

1. variations in oxygen concentration within a sealed bottle (developed by Gaarder and Gran in 1927).
2. incorporation of inorganic carbon-14 (14C in the form of sodium bicarbonate) into organic matter.
3. Stable isotopes of Oxygen (16O, 18O and 17O).
4. fluorescence kinetics (technique still a research topic).
5. Stable isotopes of Carbon (12C and 13C).
6. Oxygen/Argon Ratios.

The technique developed by Gaarder and Gran uses variations in the concentration of oxygen under different experimental conditions to infer gross primary production. Typically, three identical transparent vessels are filled with sample water and stoppered. The first is analysed immediately and used to determine the initial oxygen concentration; usually this is done by performing a Winkler titration. The other two vessels are incubated, one each in under light and darkened. After a fixed period of time, the experiment ends, and the oxygen concentration in both vessels is measured. As photosynthesis has not taken place in the dark vessel, it provides a measure of ecosystem respiration. The light vessel permits both photosynthesis and respiration, so provides a measure of net photosynthesis (i.e. oxygen production via photosynthesis subtract oxygen consumption by respiration). Gross primary production is then obtained by adding oxygen consumption in the dark vessel to net oxygen production in the light vessel.

The technique of using ^{14}C incorporation (added as labelled Na_2CO_3) to infer primary production is most commonly used today because it is sensitive, and can be used in all ocean environments. As ^{14}C is radioactive (via beta decay), it is relatively straightforward to measure its incorporation in organic material using devices such as scintillation counters.

Depending upon the incubation time chosen, net or gross primary production can be estimated. Gross primary production is best estimated using relatively short incubation times (1 hour or less), since the loss of incorporated ^{14}C (by respiration and organic material excretion / exudation) will be more limited. Net primary production is the fraction of gross production remaining after these loss processes have consumed some of the fixed carbon.

Loss processes can range between 10-60% of incorporated ^{14}C according to the incubation period, ambient environmental conditions (especially temperature) and the experimental species used. Aside from those caused by the physiology of the experimental subject itself, potential losses due to the activity of consumers also need to be considered. This is particularly true in experiments making use of natural assemblages of microscopic autotrophs, where it is not possible to isolate them from their consumers.

The methods based on stable isotopes and O_2/Ar ratios have the advantage of providing estimates of respiration rates in the light without the need of incubations in the dark. Among them, the method of the triple oxygen isotopes and O_2/Ar have the additional advantage of not needing incubations in closed containers and O_2/Ar can even be measured continuously at sea using equilibrator inlet mass spectrometry (EIMS) or a membrane inlet mass spectrometry (MIMS). However, if results relevant to the carbon cycle are desired, it is probably better to rely on methods based on carbon (and not oxygen) isotopes. It is important to notice that the method based on carbon stable isotopes is not simply an adaptation of the classic ^{14}C method, but an entirely different approach that does not suffer from the problem of lack of account of carbon recycling during photosynthesis.

Global

As primary production in the biosphere is an important part of the carbon cycle, estimating it at the global scale is important in Earth system science. However, quantifying primary production at this scale is difficult because of the range of habitats on Earth, and because of the impact of weather events (availability of sunlight, water) on its variability. Using satellite-derived estimates of the Normalized Difference Vegetation Index (NDVI) for terrestrial habitats and sea-surface chlorophyll for the oceans, it is estimated that the total (photoautotrophic) primary production for the Earth was 104.9 Gt C yr^{-1}. Of this, 56.4 Gt C yr^{-1} (53.8%), was the product of terrestrial organisms, while the remaining 48.5 Gt C yr^{-1}, was accounted for by oceanic production.

Scaling ecosystem-level GPP estimations based on eddy covariance measurements of net ecosystem exchange to regional and global values using spatial details of different predictor variables, such as climate variables and remotely sensed fAPAR or LAI led to a terrestrial gross primary production of 123±8 Gt carbon (NOT carbon dioxide) per year during 1998-2005.

In areal terms, it was estimated that land production was approximately 426 g C m^{-2} yr^{-1} (excluding areas with permanent ice cover), while that for the oceans was 140 g C m^{-2} yr^{-1}. Another significant difference between the land and the oceans lies in their standing stocks - while accounting for almost half of total production, oceanic autotrophs only account for about 0.2% of the total biomass.

Estimates

Primary productivity can be estimated by a variety of proxies. One that has particular relevance to the geological record is Barium, whose concentration in marine sediments rises in line with primary productivity at the surface.

Primary Production and Plant Biomass for the Earth

Ecosystem type	Area (10^6 km^2)	Mean NPP ($g/m^2/yr$)	World NPP (10^9 tons/yr)	Mean biomass (kg/m^2)	World biomass (10^9 tons)
Tropical rainforest	17.0	2,200	37.4	45	763
Tropical seasonal forest	7.5	1,600	12.0	35	260
Temperate evergreen forest	5.0	1,300	6.5	35	175
Temperate deciduous forest	7.0	1,200	8.4	30	210
Boreal forest	12.0	800	9.6	20	240
Woodlands and shrublands	8.5	700	6.0	6	50
Savanna	15.0	900	13.6	4	60
Temperate grasslands	9.0	600	5.4	1.6	14
Tundra and alpine	8.0	140	1.1	0.6	5
Desert and semi-desert	18.0	90	1.6	0.7	13
Extreme desert and ice	24.0	3	0.07	0.02	0.5
Cultivated land	14.0	650	9.1	1.0	14
Swamp and wetland	2.0	2,000	4.0	12.3	30
Lakes and streams	2.0	250	0.5	0.02	0.05
Total Continental	149	773	115	12.3	1837
Open ocean	332.0	125	41.5	0.003	1.0
Upwelling zones	0.4	500	0.2	0.02	0.008
Continental shelf	26.6	360	9.6	0.01	0.27
Algal bed and reef	0.6	2,500	1.6	2.0	1.2
Estuaries	1.4	1,500	2.1	1.0	1.4

Total marine	361	152	55.0	0.01	3.9
Grand total	510	333	170	3.6	1841

From R.H. Whittaker, quoted in Peter Stiling (1996), "Ecology: Theories and Applications" (Prentice Hall).

Primary Producers

Primary producers form the bases of food webs and entire ecosystems. They produce their own food either via photosynthesis or chemosynthesis, depending on the availability of sunlight. Primary producers provide food for primary consumers or herbivores, which in turn provide food for carnivores, consumers at secondary or tertiary trophic levels.

The Role of Primary Producers

Primary producers form the basis of a food web or tropic levels. Whether it's green plants using photosynthesis or benthic bacterial mats, these organisms provide food for consumers. Primary consumers are herbivore animals that eat primary producers. Carnivorous secondary consumers eat primary consumers. Tertiary consumers eat the secondary consumers. Eventually, the top member of a food web will die, and its body will feed decomposers. These decomposers provide organic material, converted for use by the next cohort of primary consumers, and begin the cycle again. Energy passes through each trophic level. However, throughout each step of the food web, energy is lost either as heat, for life processes such as movement, or in feces and decaying remains.

Producers using Photosynthesis

Plants, algae and photosynthetic bacteria comprise primary producers that use the sun's energy for photosynthesis. Both terrestrial and aquatic primary producers perform this activity. In the water column, phytoplankton living closer to the water's surface access the sun's energy. Other primary producers in aquatic environments include seaweeds, algae, green plants, and blue-green or purple bacteria. They photosynthesize carbohydrates needed for energy. Phytoplankton comprise the foundation of ocean ecosystems, as many creatures from zooplankton to jellies to whales rely on them for food.

Producers using Chemosynthesis

Some primary producers, such as some species of bacteria, do not use the sun for energy nor for fixing carbon dioxide. This process, chemosynthesis, occurs in the absence of sunlight, such as deep in the ocean around hydrothermal vents or cold water methane seeps. These primary producers metabolize inorganic material such as nitrogen, sulfur or iron to make food from carbon dioxide. These types of producers are called chemoautotrophs, and they form the basis of food webs for octopi, crabs and other sea animals. Chemoautotrophs can also be found in other hydrogen sulfide-rich ecosystems such as salt marshes.

Interconnected Success

The quality of primary producers directly affects the success of secondary consumers, and as a result, the entire food chain. Because of the interconnected nature of food webs, it is crucial for primary producers to remain intact. Even some of the smallest primary producers, such as phytoplankton, possess far-reaching effects. Many aquatic food webs rely on phytoplankton as primary producers. The different biochemical makeup of phytoplankton affects their nutritional value for zooplankton that feed on them. Biodiversity of primary producers, whether aquatic or terrestrial, yields greater success for herbivores because the animals have more nutritious options available.

Primary producers also rely on the work of decomposers to survive. Decomposers fix nitrogen so that it is available for primary producers. Various organisms work the soil to decompose organic matter and enhance its nutrients so that plants can grow and provide the basis of a food web. While different organisms can take the place of those that grow scarce, maintaining and protecting primary producers yields sustainable ecosystems.

Food Web

Food web is an important ecological concept. Basically, food web represents feeding relationships within a community (Smith and Smith 2009). It also implies the transfer of food energy from its source in plants through herbivores to carnivores (Krebs 2009). Normally, food webs consist of a number of food chains meshed together. Each food chain is a descriptive diagram including a series of arrows, each pointing from one species to another, representing the flow of food energy from one feeding group of organisms to another.

There are two types of food chains: the grazing food chain, beginning with autotrophs, and the detrital food chain, beginning with dead organic matter (Smith & Smith 2009). In a grazing food chain, energy and nutrients move from plants to the herbivores consuming them, and to the carnivores or omnivores preying upon the herbivores. In a detrital food chain, dead organic matter of plants and animals is broken down by decomposers, e.g., bacteria and fungi, and moves to detritivores and then carnivores.

Food web offers an important tool for investigating the ecological interactions that define energy flows and predator-prey relationship (Cain et al. 2008). Figure shows a simplified food web in a desert ecosystem. In this food web, grasshoppers feed on plants; scorpions prey on grasshoppers; kit foxes prey on scorpions. While the food web showed here is a simple one, most feed webs are complex and involve many species with both strong and weak interactions among them (Pimm et al. 1991). For example, the predators of a scorpion in a desert ecosystem might be a golden eagle, an owl, a roadrunner, or a fox.

The idea to apply the food chains to ecology and to analyze its consequences was first proposed by Charles Elton (Krebs 2009). In 1927, he recognized that the length of these food chains was mostly limited to 4 or 5 links and the food chains were not isolated, but hooked together into food webs (which he called "food cycles").

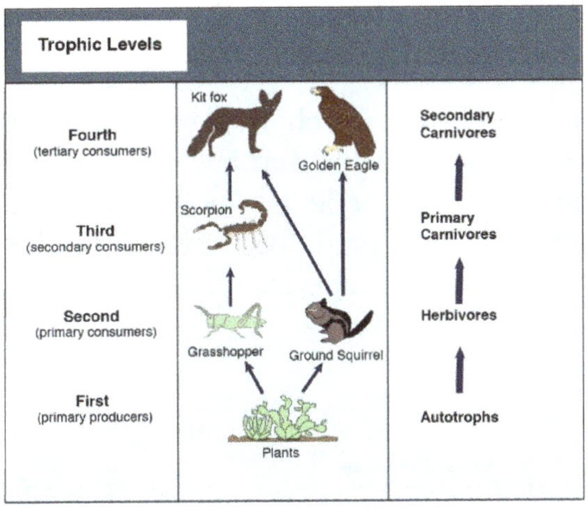

Figure: A simple six-member food web for a representative desert grassland.

Taxonomy of a Food Web

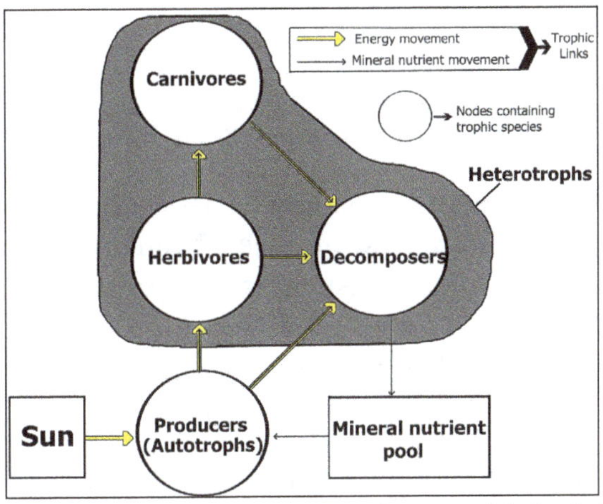

A simplified food web illustrating a three trophic food chain (producers-herbivores-carnivores) linked to decomposers.

The movement of mineral nutrients is cyclic, whereas the movement of energy is unidirectional and noncyclic. Trophic species are encircled as nodes and arrows depict the links.

Links in food webs map the feeding connections (who eats whom) in an ecological community. Food cycle is an obsolete term that is synonymous with food web. Ecologists can broadly group all life forms into one of two trophic layers, the autotrophs and the heterotrophs. Autotrophs produce more biomass energy, either chemically without the sun's energy or by capturing the sun's energy in photosynthesis, than they use during metabolic respiration. Heterotrophs consume rather than produce biomass energy as they metabolize, grow, and add to levels of secondary production. A food web depicts a collection of polyphagous heterotrophic consumers that network and cycle the flow of energy and nutrients from a productive base of self-feeding autotrophs.

The base or basal species in a food web are those species without prey and can include autotrophs or saprophytic detritivores (i.e., the community of decomposers in soil, biofilms, and periphyton). Feeding connections in the web are called trophic links. The number of trophic links per consumer is a measure of food web connectance. Food chains are nested within the trophic links of food webs. Food chains are linear (noncyclic) feeding pathways that trace monophagous consumers from a base species up to the top consumer, which is usually a larger predatory carnivore.

Linkages connect to nodes in a food web, which are aggregates of biological taxa called trophic species. Trophic species are functional groups that have the same predators and prey in a food web. Common examples of an aggregated node in a food web might include parasites, microbes, decomposers, saprotrophs, consumers, or predators, each containing many species in a web that can otherwise be connected to other trophic species.

Trophic Levels

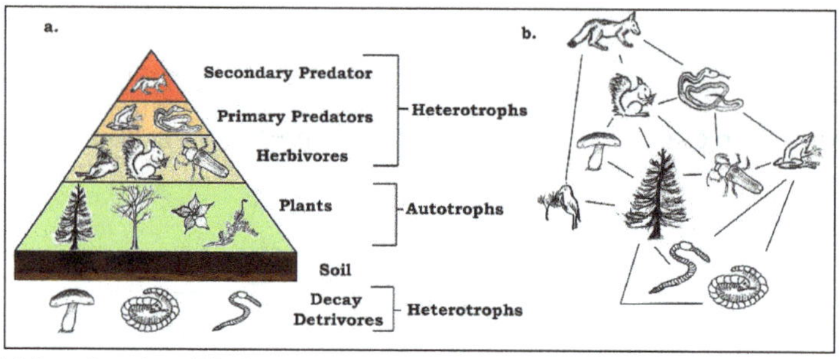

A trophic pyramid (a) and a simplified community food web (b) illustrating ecological relations among creatures that are typical of a northern Boreal terrestrial ecosystem.

The trophic pyramid roughly represents the biomass (usually measured as total dry-weight) at each level. Plants generally have the greatest biomass. Names of trophic categories are shown to the right of the pyramid. Some ecosystems, such as many wetlands, do not organize as a strict pyramid, because aquatic plants are not as productive as long-lived terrestrial plants such as trees. Ecological trophic pyramids are typically one of three kinds: 1) pyramid of numbers, 2) pyramid of biomass, or 3) pyramid of energy.

Food webs have trophic levels and positions. Basal species, such as plants, form the first level and are the resource limited species that feed on no other living creature in the web. Basal species can be autotrophs or detritivores, including "decomposing organic material and its associated microorganisms which we defined as detritus, micro-inorganic material and associated microorganisms (MIP), and vascular plant material." Most autotrophs capture the sun's energy in chlorophyll, but some autotrophs (the chemolithotrophs) obtain energy by the chemical oxidation of inorganic compounds and can grow in dark environments, such as the sulfur bacterium Thiobacillus, which lives in hot sulfur springs. The top level has top (or apex) predators which no other species kills directly for its food resource needs. The intermediate levels are filled with omnivores that feed on more than one trophic level and cause energy to flow through a number of food pathways starting from a basal species.

In the simplest scheme, the first trophic level (level 1) is plants, then herbivores (level 2), and then carnivores (level 3). The trophic level is equal to one more than the chain length, which is the number of links connecting to the base. The base of the food chain (primary producers or detritivores) is set at zero. Ecologists identify feeding relations and organize species into trophic species through extensive gut content analysis of different species. The technique has been improved through the use of stable isotopes to better trace energy flow through the web. It was once thought that omnivory was rare, but recent evidence suggests otherwise. This realization has made trophic classifications more complex.

Trophic Dynamics

The trophic level concept was introduced in a historical landmark paper on trophic dynamics in 1942 by Raymond L. Lindeman. The basis of trophic dynamics is the transfer of energy from one part of the ecosystem to another. The trophic dynamic concept has served as a useful quantitative heuristic, but it has several major limitations including the precision by which an organism can be allocated to a specific trophic level. Omnivores, for example, are not restricted to any single level. Nonetheless, recent research has found that discrete trophic levels do exist, but "above the herbivore trophic level, food webs are better characterized as a tangled web of omnivores."

A central question in the trophic dynamic literature is the nature of control and regulation over resources and production. Ecologists use simplified one trophic position food chain models (producer, carnivore, decomposer). Using these models, ecologists have tested various types of ecological control mechanisms. For example, herbivores generally have an abundance of vegetative resources, which meant that their populations were largely controlled or regulated by predators. This is known as the top-down hypothesis or 'green-world' hypothesis. Alternatively to the top-down hypothesis, not all plant material is edible and the nutritional quality or antiherbivore defenses of plants (structural and chemical) suggests a bottom-up form of regulation or control. Recent studies have concluded that both "top-down" and "bottom-up" forces can influence community structure and the strength of the influence is environmentally context dependent. These complex multitrophic interactions involve more than two trophic levels in a food web.

Another example of a multi-trophic interaction is a trophic cascade, in which predators help to increase plant growth and prevent overgrazing by suppressing herbivores. Links in a food-web illustrate direct trophic relations among species, but there are also indirect effects that can alter the abundance, distribution, or biomass in the trophic levels. For example, predators eating herbivores indirectly influence the control and regulation of primary production in plants. Although the predators do not eat the plants directly, they regulate the population of herbivores that are directly linked to plant trophism. The net effect of direct and indirect relations is called trophic cascades. Trophic cascades are separated into species-level cascades, where only a subset of the food-web dynamic is impacted by a change in population numbers, and community-level cascades, where a change in population numbers has a dramatic effect on the entire food-web, such as the distribution of plant biomass.

Energy Flow and Biomass

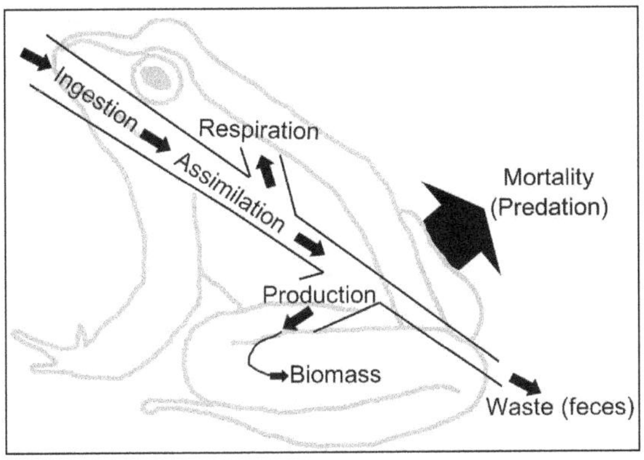

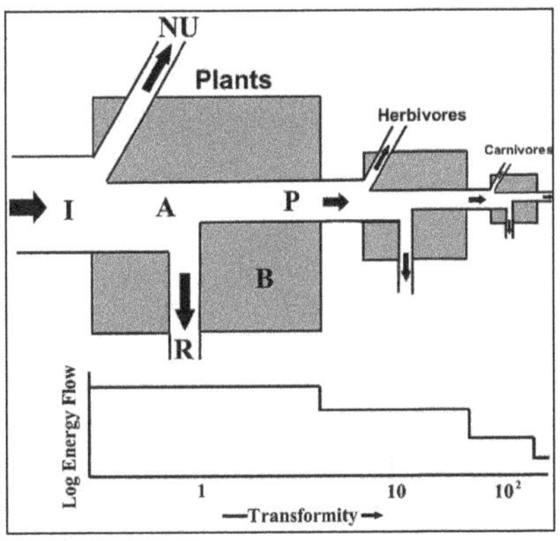

Left: Energy flow diagram of a frog. The frog represents a node in an extended food web. The energy ingested is utilized for metabolic processes and transformed into biomass. The energy flow continues on its path if the frog is ingested by predators, parasites, or as a decaying carcass in soil. This energy flow diagram illustrates how energy is lost as it fuels the metabolic process that transform the energy and nutrients into biomass.

Right: An expanded three link energy food chain (1. plants, 2. herbivores, 3. carnivores) illustrating the relationship between food flow diagrams and energy transformity. The transformity of energy becomes degraded, dispersed, and diminished from higher quality to lesser quantity as the energy within a food chain flows from one trophic species into another. Abbreviations: I=input, A=assimilation, R=respiration, NU=not utilized, P=production, B=biomass.

Food webs depict energy flow via trophic linkages. Energy flow is directional, which contrasts against the cyclic flows of material through the food web systems. Energy flow "typically includes production, consumption, assimilation, non-assimilation losses (feces), and respiration (maintenance costs)." In a very general sense, energy flow (E) can be defined as the sum of metabolic production (P) and respiration (R), such that E=P+R.

Biomass represents stored energy. However, concentration and quality of nutrients and energy is variable. Many plant fibers, for example, are indigestible to many herbivores leaving grazer community food webs more nutrient limited than detrital food webs where bacteria are able to access and release the nutrient and energy stores. "Organisms usually extract energy in the form of carbohydrates, lipids, and proteins. These polymers have a dual role as supplies of energy as well as building blocks; the part that functions as energy supply results in the production of nutrients (and carbon dioxide, water, and heat). Excretion of nutrients is, therefore, basic to metabolism." The units in energy flow webs are typically a measure mass or energy per m^2 per unit time. Different consumers are going to have different metabolic assimilation efficiencies in their diets. Each trophic level transforms energy into biomass. Energy flow diagrams illustrate the rates and efficiency of transfer from one trophic level into another and up through the hierarchy.

It is the case that the biomass of each trophic level decreases from the base of the chain to the top. This is because energy is lost to the environment with each transfer as entropy increases. About eighty to ninety percent of the energy is expended for the organism's life processes or is lost as heat or waste. Only about ten to twenty percent of the organism's energy is generally passed to the next organism. The amount can be less than one percent in animals consuming less digestible plants, and it can be as high as forty percent in zooplankton consuming phytoplankton. Graphic representations of the biomass or productivity at each tropic level are called ecological pyramids or trophic pyramids. The transfer of energy from primary producers to top consumers can also be characterized by energy flow diagrams.

Food Chain

A common metric used to quantify food web trophic structure is food chain length. Food chain length is another way of describing food webs as a measure of the number of species encountered as energy or nutrients move from the plants to top predators. There are different ways of calculating food chain length depending on what parameters of the food web dynamic are being considered: connectance, energy, or interaction. In its simplest form, the length of a chain is the number of links between a trophic consumer and the base of the web. The mean chain length of an entire web is the arithmetic average of the lengths of all chains in a food web.

In a simple predator-prey example, a deer is one step removed from the plants it eats (chain length = 1) and a wolf that eats the deer is two steps removed from the plants (chain length = 2). The relative amount or strength of influence that these parameters have on the food web address questions about:

- the identity or existence of a few dominant species (called strong interactors or keystone species).
- the total number of species and food-chain length (including many weak interactors) and
- how community structure, function and stability is determined.

Ecological Pyramids

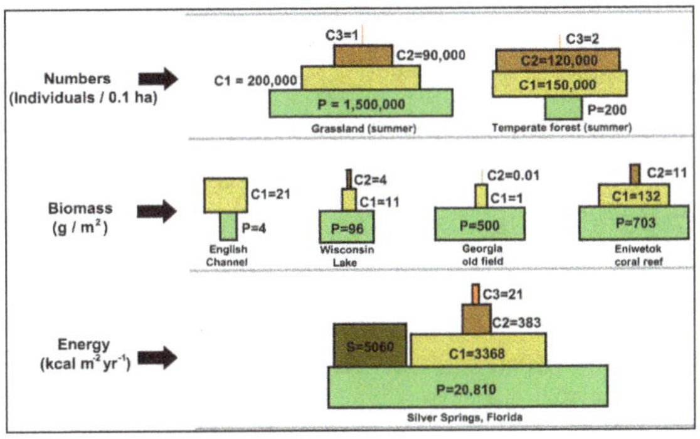

Top Left: A four level trophic pyramid sitting on a layer of soil and its community of decomposers. Top right: A three layer trophic pyramid linked to the biomass and energy flow concepts. Bottom: Illustration of a range of ecological pyramids, including top pyramid of numbers, middle pyramid of biomass, and bottom pyramid of energy. The terrestrial forest (summer) and the English Channel ecosystems exhibit inverted pyramids.Note: trophic levels are not drawn to scale and the pyramid of numbers excludes microorganisms and soil animals. Abbreviations: P=Producers, C1=Primary consumers, C2=Secondary consumers, C3=Tertiary consumers, S=Saprotrophs.

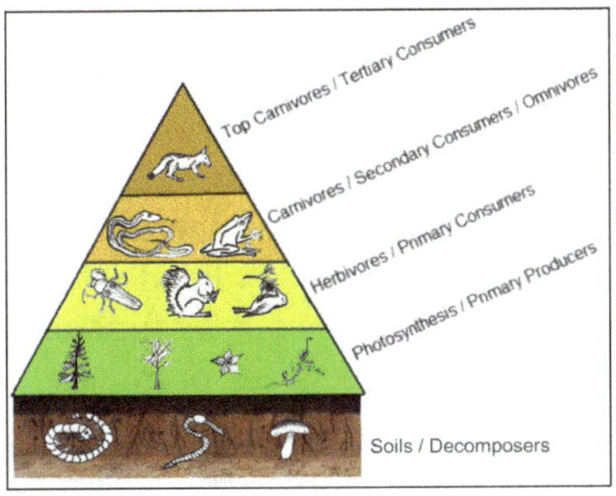

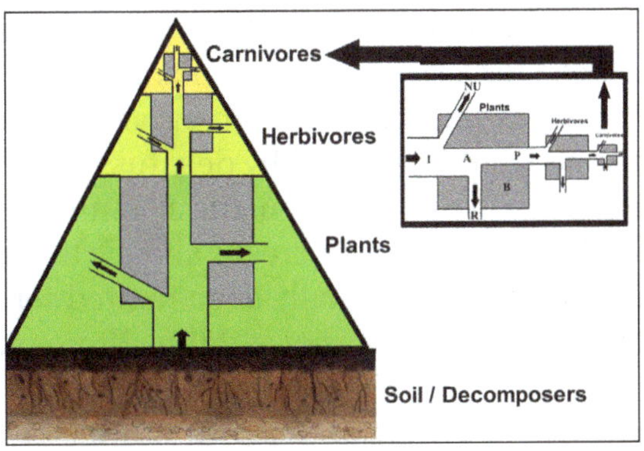

In a pyramid of numbers, the number of consumers at each level decreases significantly, so that a single top consumer, (e.g., a polar bear or a human), will be supported by a much larger number of separate producers. There is usually a maximum of four or five links in a food chain, although food chains in aquatic ecosystems are more often longer than those on land. Eventually, all the energy in a food chain is dispersed as heat.

Ecological pyramids place the primary producers at the base. They can depict different numerical properties of ecosystems, including numbers of individuals per unit of area, biomass (g/m^2), and energy (k cal m^{-2} yr^{-1}). The emergent pyramidal arrangement of trophic levels with amounts of energy transfer decreasing as species become further removed from the source of production is one of several patterns that is repeated amongst the planets ecosystems.\ The size of each level in the pyramid generally represents biomass, which can be measured as the dry weight of an organism. Autotrophs may have the highest global proportion of biomass, but they are closely rivaled or surpassed by microbes.

Pyramid structure can vary across ecosystems and across time. In some instances biomass pyramids can be inverted. This pattern is often identified in aquatic and coral reef ecosystems. The pattern of biomass inversion is attributed to different sizes of producers. Aquatic communities are often dominated by producers that are smaller than the consumers that have high growth rates. Aquatic producers, such as planktonic algae or aquatic plants, lack the large accumulation of secondary growth as exists in the woody trees of terrestrial ecosystems. However, they are able to reproduce quickly enough to support a larger biomass of grazers. This inverts the pyramid. Primary consumers have longer lifespans and slower growth rates that accumulates more biomass than the producers they consume. Phytoplankton live just a few days, whereas the zooplankton eating the phytoplankton live for several weeks and the fish eating the zooplankton live for several consecutive years. Aquatic predators also tend to have a lower death rate than the smaller consumers, which contributes to the inverted pyramidal pattern. Population structure, migration rates, and environmental refuge for prey are other possible causes for pyramids with biomass inverted. Energy pyramids, however, will always have an upright pyramid shape if all sources of food energy are included and this is dictated by the second law of thermodynamics.

Material Flux and Recycling

Many of the Earth's elements and minerals (or mineral nutrients) are contained within the tissues and diets of organisms. Hence, mineral and nutrient cycles trace food web energy pathways. Ecologists employ stoichiometry to analyze the ratios of the main elements found in all organisms: carbon (C), nitrogen (N), phosphorus (P). There is a large transitional difference between many terrestrial and aquatic systems as C:P and C:N ratios are much higher in terrestrial systems while N:P ratios are equal between the two systems. Mineral nutrients are the material resources that organisms need for growth, development, and vitality. Food webs depict the pathways of mineral nutrient cycling as they flow through organisms. Most of the primary production in an ecosystem is not consumed, but is recycled by detritus back into useful nutrients. Many of the Earth's microorganisms are involved in the formation of minerals in a process called biomineralization. Bacteria that live in detrital sediments create and cycle nutrients and biominerals. Food web models and nutrient cycles have traditionally been treated separately, but there is a strong functional connection between the two in terms of stability, flux, sources, sinks, and recycling of mineral nutrients.

Kinds of Food Webs

Food webs are necessarily aggregated and only illustrate a tiny portion of the complexity of real ecosystems. For example, the number of species on the planet are likely in the general order of

10^7, over 95% of these species consist of microbes and invertebrates, and relatively few have been named or classified by taxonomists. It is explicitly understood that natural systems are 'sloppy' and that food web trophic positions simplify the complexity of real systems that sometimes overemphasize many rare interactions. Most studies focus on the larger influences where the bulk of energy transfer occurs. "These omissions and problems are causes for concern, but on present evidence do not present insurmountable difficulties."

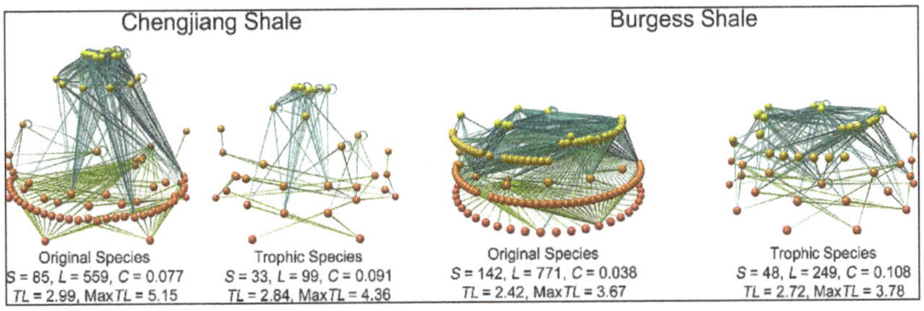

Paleoecological studies can reconstruct fossil food-webs and trophic levels. Primary producers form the base (red spheres), predators at top (yellow spheres), the lines represent feeding links. Original food-webs (left) are simplified (right) by aggregating groups feeding on common prey into coarser grained trophic species.

There are Different Kinds or Categories of Food Webs:

- Source web - one or more node(s), all of their predators, all the food these predators eat, and so on.
- Sink web - one or more node(s), all of their prey, all the food that these prey eat, and so on.
- Community (or connectedness) web - a group of nodes and all the connections of who eats whom.
- Energy flow web - quantified fluxes of energy between nodes along links between a resource and a consumer.
- Paleoecological web - a web that reconstructs ecosystems from the fossil record.
- Functional web - emphasizes the functional significance of certain connections having strong interaction strength and greater bearing on community organization, more so than energy flow pathways. Functional webs have compartments, which are sub-groups in the larger network where there are different densities and strengths of interaction. Functional webs emphasize that "the importance of each population in maintaining the integrity of a community is reflected in its influence on the growth rates of other populations."

Within these categories, food webs can be further organized according to the different kinds of ecosystems being investigated. For example, human food webs, agricultural food webs, detrital food webs, marine food webs, aquatic food webs, soil food webs, Arctic (or polar) food webs, terrestrial food webs, and microbial food webs. These characterizations stem from the ecosystem concept, which assumes that the phenomena under investigation (interactions and feedback loops) are sufficient to explain patterns within boundaries, such as the edge of a forest, an island, a shoreline, or some other pronounced physical characteristic.

Detrital Web

In a detrital web, plant and animal matter is broken down by decomposers, e.g., bacteria and fungi, and moves to detritivores and then carnivores. There are often relationships between the detrital web and the grazing web. Mushrooms produced by decomposers in the detrital web become a food source for deer, squirrels, and mice in the grazing web. Earthworms eaten by robins are detritivores consuming decaying leaves.

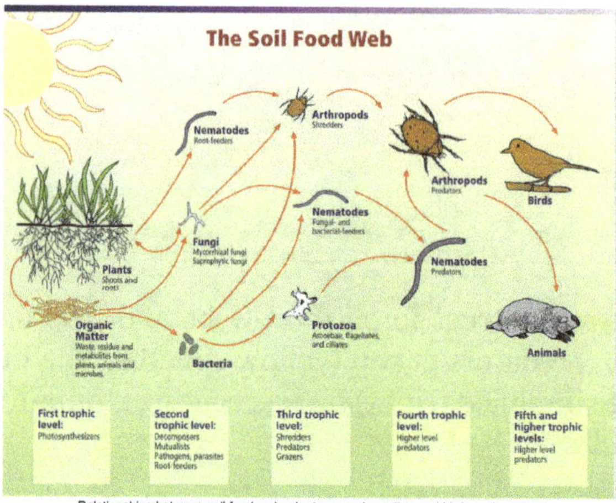

Relationships between soil food web, plants, organic matter, and birds and mammals
Image courtesy of USDA Natural Resources Conservation Service
http://soils.usda.gov/sqi/soil_quality/soil_biology/soil_food_web.html.

An illustration of a soil food web.

"Detritus can be broadly defined as any form of non-living organic matter, including different types of plant tissue (e.g. leaf litter, dead wood, aquatic macrophytes, algae), animal tissue (carrion), dead microbes, faeces (manure, dung, faecal pellets, guano, frass), as well as products secreted, excreted or exuded from organisms (e.g. extra-cellular polymers, nectar, root exudates and leachates, dissolved organic matter, extra-cellular matrix, mucilage). The relative importance of these forms of detritus, in terms of origin, size and chemical composition, varies across ecosystems."

Quantitative Food Webs

Ecologists collect data on trophic levels and food webs to statistically model and mathematically calculate parameters, such as those used in other kinds of network analysis (e.g., graph theory), to study emergent patterns and properties shared among ecosystems. There are different ecological dimensions that can be mapped to create more complicated food webs, including: species composition (type of species), richness (number of species), biomass (the dry weight of plants and animals), productivity (rates of conversion of energy and nutrients into growth), and stability (food webs over time). A food web diagram illustrating species composition shows how change in a single species can directly and indirectly influence many others. Microcosm studies are used to simplify food web research into semi-isolated units such as small springs, decaying logs, and laboratory experiments using organisms that reproduce quickly, such as daphnia feeding on algae grown under controlled environments in jars of water.

While the complexity of real food webs connections are difficult to decipher, ecologists have found mathematical models on networks an invaluable tool for gaining insight into the structure, stability, and laws of food web behaviours relative to observable outcomes. "Food web theory centers around the idea of connectance." Quantitative formulas simplify the complexity of food web structure. The number of trophic links (t_L), for example, is converted into a connectance value:

$$C = \frac{t_L}{S(S-1)/2},$$

where, S(S-1)/2 is the maximum number of binary connections among S species. "Connectance (C) is the fraction of all possible links that are realized (L/S^2) and represents a standard measure of food web complexity..." The distance (d) between every species pair in a web is averaged to compute the mean distance between all nodes in a web (D) and multiplied by the total number of links (L) to obtain link-density (LD), which is influenced by scale dependent variables such as species richness. These formulas are the basis for comparing and investigating the nature of non-random patterns in the structure of food web networks among many different types of ecosystems.

Scaling laws, complexity, choas, and patterned correlates are common features attributed to food web structure.

Complexity and Stability

Food webs are complex. Complexity is a measure of an increasing number of permutations and it is also a metaphorical term that conveys the mental intractability or limits concerning unlimited algorithmic possibilities. In food web terminology, complexity is a product of the number of species and connectance. Connectance is "the fraction of all possible links that are realized in a network".These concepts were derived and stimulated through the suggestion that complexity leads to stability in food webs, such as increasing the number of trophic levels in more species rich ecosystems. This hypothesis was challenged through mathematical models suggesting otherwise, but subsequent studies have shown that the premise holds in real systems.

At different levels in the hierarchy of life, such as the stability of a food web, "the same overall structure is maintained in spite of an ongoing flow and change of components." The farther a living system (e.g., ecosystem) sways from equilibrium, the greater its complexity. Complexity has multiple meanings in the life sciences and in the public sphere that confuse its application as a precise term for analytical purposes in science. Complexity in the life sciences (or biocomplexity) is defined by the "properties emerging from the interplay of behavioral, biological, physical, and social interactions that affect, sustain, or are modified by living organisms, including humans".

Several concepts have emerged from the study of complexity in food webs. Complexity explains many principals pertaining to self-organization, non-linearity, interaction, cybernetic feedback, discontinuity, emergence, and stability in food webs. Nestedness, for example, is defined as "a pattern of interaction in which specialists interact with species that form perfect subsets of the species with which generalists interact", "—that is, the diet of the most specialized species is a subset of the diet of the next more generalized species, and its diet a subset of the next more generalized, and so on." Until recently, it was thought that food webs had little nested structure, but empirical evidence shows that many published webs have nested subwebs in their assembly.

Food webs are complex networks. As networks, they exhibit similar structural properties and mathematical laws that have been used to describe other complex systems, such as small world and scale free properties. The small world attribute refers to the many loosely connected nodes, non-random dense clustering of a few nodes (i.e., trophic or keystone species in ecology), and small path length compared to a regular lattice. "Ecological networks, especially mutualistic networks, are generally very heterogeneous, consisting of areas with sparse links among species and distinct areas of tightly linked species. These regions of high link density are often referred to as cliques, hubs, compartments, cohesive sub-groups, or modules...Within food webs, especially in aquatic systems, nestedness appears to be related to body size because the diets of smaller predators tend to be nested subsets of those of larger predators (Woodward & Warren 2007; YvonDurocher et al. 2008), and phylogenetic constraints, whereby related taxa are nested based on their common evolutionary history, are also evident (Cattin et al. 2004)." "Compartments in food webs are subgroups of taxa in which many strong interactions occur within the subgroups and few weak interactions occur between the subgroups. Theoretically, compartments increase the stability in networks, such as food webs."

Food webs are also complex in the way that they change in scale, seasonally, and geographically. The components of food webs, including organisms and mineral nutrients, cross the thresholds of ecosystem boundaries. This has led to the concept or area of study known as cross-boundary subsidy. "This leads to anomalies, such as food web calculations determining that an ecosystem can support one half of a top carnivore, without specifying which end." Nonetheless, real differences in structure and function have been identified when comparing different kinds of ecological food webs, such as terrestrial vs. aquatic food webs.

Applications of Food Webs

Food webs are constructed to describe species interactions (direct relationships).

The fundamental purpose of food webs is to describe feeding relationship among species in a community. Food webs can be constructed to describe the species interactions. All species in the food webs can be distinguished into basal species (autotrophs, such as plants), intermediate species (herbivores and intermediate level carnivores, such as grasshopper and scorpion) or top predators (high level carnivores such as fox).

These feeding groups are referred as trophic levels. Basal species occupy the lowest trophic level as primary producer. They convert inorganic chemical and use solar energy to generate chemical energy. The second trophic level consists of herbivores. These are first consumers. The remaining trophic levels include carnivores that consume animals at trophic levels below them. The second consumers (trophic level 3) in the desert food web include birds and scorpions, and tertiary consumers making up the fourth trophic level include bird predators and foxes. Grouping all species into different functional groups or tropic levels helps us simplify and understand the relationships among these species.

Food webs can be used to illustrate indirect interactions among species.

Indirect interaction occurs when two species do not interact with each other directly, but influenced by a third species. Species can influence one another in many different ways. One example is the keystone predation are demonstrated by Robert Paine in an experiment conducted in the rocky intertidal zone (Cain et al. 2008; Smith & Smith 2009; Molles 2010). This study showed that predation can influence the competition among species in a food web. The intertidal zone is home to a variety of mussels, barnacles, limpets, and chitons (Paine 1969). All these invertebrate herbivores are preyed upon by the predator starfish Pisaster. Starfish was relatively uncommon in the intertidal zone, and considered less important in the community. When Paine manually removed the starfish from experimental plots while leaving other areas undisturbed as control plots, he found that the number of prey species in the experimental plots dropped from 15 at the beginning of the experiment to 8 (a loss of 7 species) two years after the starfish removal while the total of prey species remained the same in the control plots. He reasoned that in the absence of the predator starfish, several of the mussel and barnacle species (that were superior competitors) excluded the other species and reduced overall diversity in the community (Smith & Smith 2009). Predation by starfish reduced the abundance of mussel and opened up space for other species to colonize and persist. This type of indirect interaction is called keystone predation.

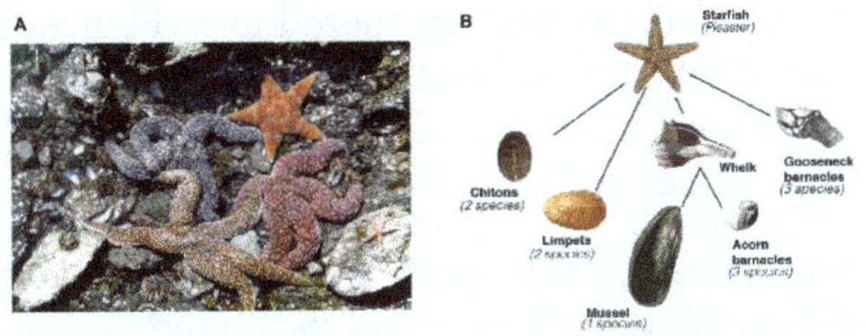

(a) The rocky intertidal zone of the Pacific Northwest coast is inhabited by a variety of species including starfish, barnacles, limpets, chitons, and mussels. (b) A food web of this community shows that the starfish preys on a variety of invertebrate species.

Another interesting study demonstrated indirect interactions among species in both aquatic and terrestrial ecosystems. In a study conducted near Gainesville, Florida, Knight and her colleagues (2009) investigated the effects of fish in ponds on plant seeds production. They measured and compared abundances of both larval and adult dragonfly in and around four ponds that had been stocked with fish and four ponds that lacked fish (Knight et al. 2009). They found that ponds with fish produce fewer larval and adult dragonflies than ponds without fish, as fish prey on larval dragonflies. As dragonfly population decreases, the populations of their prey, including bees, flies, and butterflies, decrease. These prey species are pollinators of the plants. Therefore, flowers in the vicinity of ponds without fish receive fewer pollinator visits than flowers close to ponds stocked with fish. Since the production of seeds is pollen-limited, fewer pollinator visits result in lower seeds production. This study demonstrates, via a complex trophic cascade, that adding fish to a pond improves the reproductive success of a plant on land (Ricklefs 2008).

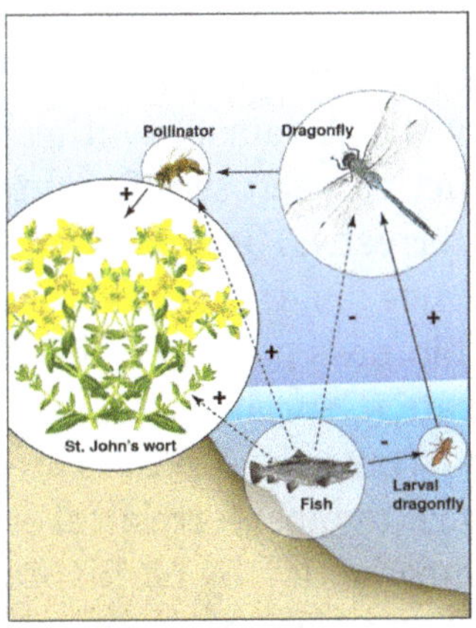

An interaction food web shows that fish have indirect effects on the populations of several species in and around ponds.

The solid arrows represent direct effects, and the dashed arrows indirect effects; the nature of the effect is indicated by + or -. Fish have indirect effects, through a trophic cascade, on several terrestrial species: dragonfly adults (-), pollinators (+), and plants (+)

Food webs can be used to study bottom-up or top-down control of community structure.

Food webs illustrate energy flow from primary producers to primary consumers (herbivores), and from primary consumers to secondary consumers (carnivores). The structure of food webs suggests that productivity and abundance of populations at any given trophic level are controlled by the productivity and abundance of populations in the trophic level below them (Smith & Smith 2009). This phenomenon is call bottom-up control. Correlations in abundance or productivity between consumers and their resources are considered as evidence for bottom-up control. For example, plant population densities control the abundance of herbivore populations which in turn control the densities of the carnivore populations. Thus, the biomass of herbivores usually increases with primary productivity in terrestrial ecosystems.

Top-down control occurs when the population density of a consumer can control that of its resource, for example, predator populations can control the abundance of prey species (Power 1992). Under top-down control, the abundance or biomass of lower trophic levels depends on effects from consumers at higher trophic levels. A trophic cascade is a type of top-down interaction that describes the indirect effects of predators. In a trophic cascade, predators induce effects that cascade down the food chain and affect biomass of organisms at least two links away (Ricklefs 2008). Nelson Hairston, Frederick Smith and Larry Slobodkin first introduced the concept of top-down control with the frequently quoted "the world is green" proposition (Power 1992; Smith & Smith 2009). They proposed that the world is green because carnivores depress herbivores and keep herbivore populations in check. Otherwise, herbivores would consume most of the vegetation. Indeed, a bird exclusion study demonstrated that there were significantly more insects and leaf damage in plots without birds compared to the control (Marquis & Whelan 1994).

Food webs can be used to reveal different patterns of energy transfer in terrestrial and aquatic ecosystems.

Patterns of energy flow through different ecosystems may differ markedly in terrestrial and aquatic ecosystems (Shurin et al. 2006). Food webs (i.e., energy flow webs) can be used to reveal these differences. In a review paper, Shurin et al. (2006) provided evidence for systematic difference in energy flow and biomass partitioning between producers and herbivores, detritus and decomposers, and higher trophic levels in food webs. A dataset synthesized by Cebrian and colleagues on the fate of carbon fixed by primary productivity across different ecosystems was used to show different patterns in food chains between terrestrial and aquatic ecosystems. On average, the turnover rate of phytoplankton is 10 to 1000 times faster than that of grasslands and forests, thus, less carbon is stored in the living autotroph biomass pool, and producer biomass is consumed by aquatic herbivores at 4 times the terrestrial rate (Cebrian 1999, 2004; Shurin et al. 2006). Herbivores in terrestrial ecosystems are less abundant but decomposers are much more abundant than in phytoplankton dominated aquatic ecosystems. In most terrestrial ecosystems with high standing biomass and relatively low harvest of primary production by herbivores, the detrital food chain is dominant (Smith & Smith 2009). In deep-water aquatic ecosystems, with their low standing biomass, rapid turnover of organisms, and high rate of harvest, the grazing food chain may be dominant.

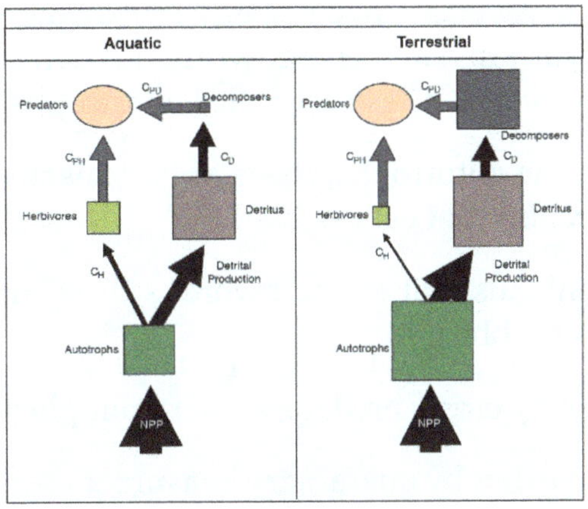

Differences in pathways of carbon flow and pools between aquatic and terrestrial ecosystems.

The thickness of the arrows (flows) and the area of the boxes (pools) correspond to the magnitude. The size of the pools are scaled as log units since the differences cover four orders of magnitude. The C's indicate consumption terms (i.e. CH is consumption by herbivores). Ovals and arrows in grey indicate unknown quantities.

As a diagram tool, food web has been approved to be effective in illustrating species interactions and testing research hypotheses. It will continue to be very helpful for us to understand the associations of species richness/diversity with food web complexity, ecosystem productivity, and stability.

Food Chain

Sample Food Chains

Trophic Level	Grassland Biome	Pond Biome	Ocean Biome
Primary Producer	grass	algae	phytoplankton
Primary Consumer	grasshopper	mosquito larva	zooplankton
Secondary Consumer	rat	dragonfly larva	fish
Tertiary Consumer	snake	fish	seal
Quaternary Consumer	hawk	raccoon	white shark

Every organism needs to obtain energy in order to live. For example, plants get energy from the sun, some animals eat plants, and some animals eat other animals.

A food chain is the sequence of who eats whom in a biological community (an ecosystem) to obtain nutrition. A food chain starts with the primary energy source, usually the sun or boiling-hot deep sea vents. The next link in the chain is an organism that make its own food from the primary energy source -- an example is photosynthetic plants that make their own food from sunlight (using a process called photosynthesis) and chemosynthetic bacteria that make their food energy from chemicals in hydrothermal vents. These are called autotrophs or primary producers.

Next come organisms that eat the autotrophs; these organisms are called herbivores or primary consumers -- an example is a rabbit that eats grass.

The next link in the chain is animals that eat herbivores - these are called secondary consumers -- an example is a snake that eat rabbits.

In turn, these animals are eaten by larger predators -- an example is an owl that eats snakes.

The tertiary consumers are are eaten by quaternary consumers -- an example is a hawk that eats owls. Each food chain end with a top predator, and animal with no natural enemies (like an alligator, hawk, or polar bear).

The arrows in a food chain show the flow of energy, from the sun or hydrothermal vent to a top predator. As the energy flows from organism to organism, energy is lost at each step. A network of many food chains is called a food web.

Ecological Pyramid

An ecological pyramid is a graphical representation of the relationship between different organisms in an ecosystem. Each of the bars that make up the pyramid represents a different trophic level, and their order, which is based on who eats whom, represents the flow of energy. Energy moves up the pyramid, starting with the primary producers, or autotrophs, such as plants and algae at the very bottom, followed by the primary consumers, which feed on these plants, then secondary consumers, which feed on the primary consumers, and so on. The height of the bars should all be the same, but the width of each bar is based on the quantity of the aspect being measured.

Types of Ecological Pyramids

Pyramid of Numbers

This shows the number of organisms in each trophic level without any consideration for their size. This type of pyramid can be convenient, as counting is often a simple task and can be done over the years to observe the changes in a particular ecosystem. However, some types of organisms are difficult to count, especially when it comes to some juvenile forms. Unit: number of organisms.

Pyramid of Biomass

This indicates the total mass of organisms at each trophic level. Usually, this type of pyramid is largest at the bottom and gets smaller going up, but exceptions do exist. The biomass of one trophic level is calculated by multiplying the number of individuals in the trophic level by the average mass of one individual in a particular area. This type of ecological pyramid solves some problems of the pyramid of numbers, as it shows a more accurate representation of the amount of energy contained in each trophic level, but it has its own limitations. For example, the time of year when the data are gathered is very important, since different species have different breeding seasons. Also, since it's usually impossible to measure the mass of every single organism, only a sample is taken, possibly leading to inaccuracies. Unit: g m^{-2} or Kg m^{-2}.

Pyramid of Productivity

The pyramid of productivity looks at the total amount of energy present at each trophic level, as well as the loss of energy between trophic levels. Since this type of representation takes into account the fact that the majority of the energy present at one trophic level will not be available for the next one, it is more accurate than the other two pyramids. This idea is based on Lindeman's Ten Percent Law, which states that only about 10% of the energy in a trophic level will go towards creating biomass. In other words, only about 10% of the energy will go into making tissue, such as stems, leaves, muscles, etc. in the next trophic level. The rest is used in respiration, hunting, and other activities, or is lost to the surroundings as heat. What's interesting, however, is that toxins are passed up the pyramid very efficiently, which means that as we go up the ecological pyramid, the amount of harmful chemicals is more and more concentrated in the organisms' bodies. This is what we call biomagnification.

The pyramid of productivity is the most widely used type of ecological pyramid, and, unlike the two other types, can never be largest at the apex and smallest at the bottom. It's an important type of ecological pyramid because it examines the flow of energy in an ecosystem over time. Unit: J m^{-2} yr^{-1}, where Joule is the unit for energy, which can be interchanged by other units of energy such as Kilojoule, Kilocalorie, and calorie.

While a productivity pyramid always takes an upright pyramid shape, number pyramids are sometimes inverted, or don't take the shape of an actual pyramid at all. To demonstrate, let's take an oak tree, which can feed millions of oakworms. If we consider this ecosystem as our focus, then the producers' level (one tree) will end up much smaller than the primary consumers' level (millions of insects). This is less likely to occur in biomass pyramids, but is not impossible. The pyramids below show the different types of pyramids and the shapes they can have in different ecosystems.

Ecological Pyramid Examples

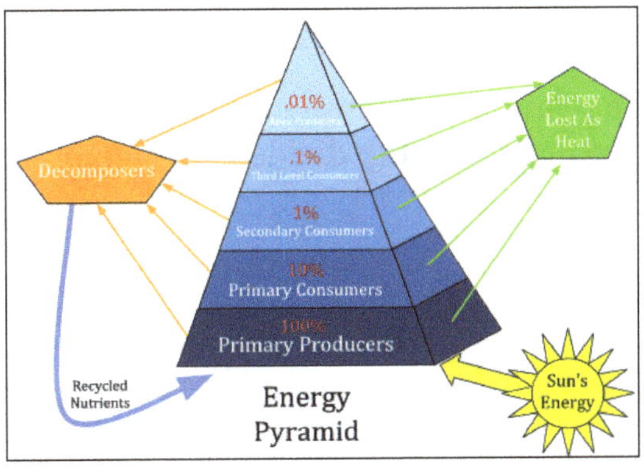

The diagram below is an example of a productivity pyramid, otherwise called an energy pyramid.

The sun has been included in this diagram, as it's the main source of all energy, as well the decomposers, like bacteria and fungi, which can acquire nutrients and energy from all trophic levels by breaking down dead or decaying organisms. As shown, the nutrients then go back into the soil and are taken up by plants.

The loss of energy to the surroundings is also shown in this diagram, and the total energy transfer has been calculated. We start off with the total amount of energy that the primary producers contain, which is indicated by 100%. As we go up one level, 90% of that energy is used in ways other than to create flesh. What the primary consumers end up with is just 10% of the starting energy, and, 10% of that 10% is lost in the transfer to the next level. That's 1%, and so on. The predators at the apex, then, will only receive 0.01% of the starting energy! This inefficiency in the system is the reason why productivity pyramids are always upright.

Function of Ecological Pyramid

An ecological pyramid not only shows us the feeding patterns of organisms in different ecosystems, but can also give us an insight into how inefficient energy transfer is, and show the influence that a change in numbers at one trophic level can have on the trophic levels above and below it. Also, when data are collected over the years, the effects of the changes that take place in the environment on the organisms can be studied by comparing the data. If an ecosystem's conditions are found to be worsening over the years because of pollution or overhunting by humans, action can be taken to prevent further damage and possibly reverse some of the present damage.

Ecological Efficiency

Ecological efficiency measures the success of production activities in minimizing a required natural flow.

A change in ecological efficiencies can shift budget shares and share limits, and thus ecological limits and target quantities.

The concept is also required to establish target rates for natural flows. Note that, although the other limits discussed are based on biological flows, ecological efficiency applies to all natural flows, and can therefore be specified for nonrenewables as well.

Ecological efficiency (E_E) is the relationship between a natural flow and its associated output, and is expressed as a ratio — the output's quantity divided by the amount of the flow used in its production:

$$E_E = Q/\text{flow}$$

If an output incorporates multiple natural flows, it will have multiple ecological efficiencies associated with it.

Because ecological efficiency is a technical measure (its definition excludes health), it is relevant to production in general, and can therefore be applied to any stage of the output life cycle.

For instance, even though a printing press is an intermediate output and thus lacks potential value, we can calculate the ecological efficiencies for the natural flows used in its production.

Ecological efficiency is a ratio of mixed dimensions. The numerator is a unit quantity of an output, and the denominator reflects the material nature of the associated flow. Following are three examples of such ratios:

- One house/Board feet of lumber.
- One construction beam/Pounds of iron.
- 1,000 hours of consulting services/Kilograms of greenhouse gases.

The mixed nature of this ratio means that ecological efficiencies are commensurable only if the outputs are the same and the natural flows are of the same types. For example, it is possible to compare the ecological efficiencies of two house-building methods by citing how many board feet of lumber each requires for a standard-size house.

However, it is not possible to compare these ecological efficie cies with those for the construction of a commercial building that uses steel.

The effects of changes in ecological efficiencies on share limits (for a single final output) and budget limits (for the economy as a whole) are depicted in the following figure.

Ecological efficiency and share/budget limits

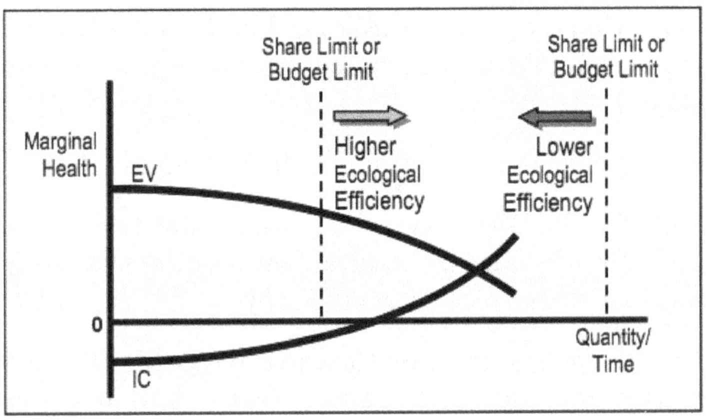

An increase in ecological efficiency will increase a share limit or budget limit, thus permitting higher output quantities. A decrease has the opposite effect. "EV" and "IC" indicate effectual value and input cost respectively.

Consider a single output first. If ecological efficiency rises for a biological flow, less of the flow is required per unit of output, which means that more of the output can be produced within its budget share.

The output's associated share limit thus increases — that is, it shifts to the right. If this share limit is also the ecological limit, the ecological limit increases. If the ecological limit is also the target quantity, the target quantity increases as well.

The reverse sequence applies if ecological efficiency falls. For an economy's total outputs, a rise in ecological efficiency means that the associated budget limit shifts to the right. If ecological efficiency falls, the budget limit shifts to the left. The impacts on the ecological limit and target scale are parallel to those for a single output.

It is important to recognize that increased ecological efficiency is the only way to increase share limits and budget limits. These limits are based entirely on the physical world, and therefore cannot be affected by economic factors such value, cost, or population levels.

On a graph the limits could therefore be imagined as fixed, with the other quantities shifting around them, unless technical changes occur that modify their associated ecological efficiencies.

Ecological efficiencies can be illustrated by the example of houses. If we stop clearcut logging and choose a more environmentally benign method of harvesting wood, habitat destruction per unit of wood will decrease.

This will constitute an increase in ecological efficiency for this flow, and will permit us to extract more wood and thus to build more houses.

If we are sufficiently successful in reducing habitat destruction, the ecological limit will shift to the wood budget. If this occurs, our efficiency efforts should be redirected — we should now focus on using less wood per house, building smaller houses, emphasizing multiple-occupancy residences over single-family houses, or perhaps moving away from wood to a construction material with less environmental impact.

Efforts such as this will continually shift the ecological limit to the next higher share limit, until the ecological limit is higher than the optimum quantity and we can rationally produce up to the output's economic limit.

A few additional points should be made regarding ecological efficiencies:

First, if an output is recycled rather than discarded, some of the natural resources it contains will be recovered. This will effectively reduce the natural flow requirement per unit of output quantity. Recycling is thus one way to increase ecological efficiency.

Second, ecological efficiency applies not only to scarce resources (those obtained through economic production, such as metals and oil), but also to resources that are widely available without production, such as air, sunlight, and water.

For example, a solar panel than uses less sunlight per unit of electricity is more ecologically efficient than one that uses more, even though sunlight is not the result of production. The same can be said for a windmill with respect to wind requirements.

The unconditional maximization of ecological efficiencies is necessary if we are to delay the onset of scarcity, thus postponing the day when painful choices must be made with respect to resource allocation.

Third, an increase in ecological efficiencies will typically lower natural cost. If less of a pollutant is generated per unit of output, the human damage from this production will likely be reduced as well.

Of course this reduced cost can easily be negated by an increase in output quantity. This is called the rebound effect or Jevons paradox, and arises when people consume more of the same output with the money they save from higher efficiencies. Although this is always a potential danger, it should not be a major issue for an ENL-driven economy.

Last, ecological efficiencies complicate the ranking of production facilities, which in ENL is based solely on potential gains. However, it is possible that facility A achieves higher potential gains than facility B, but that B has higher ecological efficiencies than A.

In such cases, which production facility should be preferred?

Unfortunately, no analytical method appears to exist here. Potential gains relate largely to present health, whereas ecological efficiency relates largely to the natural conditions for future health. Such temporally separated quantities are difficult to compare, and social judgment must therefore determine which production facility should be preferred.

ENL's definition of ecological efficiency is strongly reminiscent of the well-known "IPAT" equation. This was developed in the 1970s during debates about the factors involved in ecological impact.

Written out in full, the IPAT formula is this:

$$\text{Impact} = \text{Population} * \text{Affluence} * \text{Technology}$$

This means that the level of ecological impact is determined by population, affluence (per capita consumption), and the level of technology. If we interpret the latter as efficiency, it must appear as the denominator. The formula then becomes:

$$I = PA/T$$

Thus, ecological impact rises if population increases or per capita consumption increases; it drops if technical efficiency increases.

Compare this to ENL's formula for ecological efficiency: E_E = Q/flow. This can be rearranged as follows:

$$\text{Flow}=Q/E_E$$

The terms in these two formulas are equivalent: ecological impact is the result of natural flows, population multiplied by affluence results in output quantity, and technical efficiency is equivalent to ecological efficiency.

The formulas are thus consistent with each other, reflecting the fact that they are different approaches to the same underlying issues.

Primary Production

"Primary production" refers to energy fixed by plants. The total amount of energy fixed is usually called "gross production." A certain fraction of gross production is used in respiration by the plants; the remainder appears as new biomass or "net primary production." Thus for a single plant or a community of green plants:

Net Primary Production = Gross Production – Respiration (of Autotrophs).

Similar relationships occur in ecosystems except that the organic matter and respiration of heterotrophs must be included. The increase in total organic matter is "net ecosystem production"; respiration is the total respiration of the green plants (autotrophs) and the animal community and decay organisms (heterotrophs). Gross production is of course identical to that of the plant community. Thus for an ecosystem:

Net Ecosystem Production = Gross Production – Respiration (of Autotrophs and Heterotrophs).

References

- Tscharntke, T.; Hawkins, B., A., eds. (2002). Multitrophic Level Interactions. Cambridge: Cambridge University Press. p. 282. ISBN 0-521-79110-3
- Reichstein, Markus; Falge, Eva; Baldocchi, Dennis; Papale, Dario; Aubinet, Marc; Berbigier, Paul; et al. (2005). "On the separation of net ecosystem exchange into assimilation and ecosystem respiration: review and improved algorithm". Global Change Biology. 11 (9): 1424–1439. Bibcode:2005GCBio..11.1424R. doi:10.1111/j.1365-2486.2005.001002.x. ISSN 1354-1013
- Sigman, D.M.; Hain, M.P. (2012). "The Biological Productivity of the Ocean" (PDF). Nature Education Knowledge. 3 (6): 1–16. Retrieved 23, August 2020
- "Net Primary Productivity : Global Maps". earthobservatory.nasa.gov. 26 March 2018. Retrieved 07, March 2020
- Odum, E. P.; Barrett, G. W. (2005). Fundamentals of Ecology (5th ed.). Brooks/Cole, a part of Cengage Learning. ISBN 0-534-42066-4. Archived from the original on 2011-08-20
- Dunne, J. A.; Williams, R. J.; Martinez, N. D.; Wood, R. A.; Erwin, D. H.; Dobson, Andrew P. (2008). "Compilation and Network Analyses of Cambrian Food Webs". PLOS Biology. 6 (4): e102. doi:10.1371/journal.pbio.0060102

Chapter 4

Habitat and Biodiversity

The habitat and the physical environment along with the various living components of an ecosystem, biodiversity and its related cycles form the essential aspects of ecology. This chapter goes through each and every aspect of ecology in a detailed manner to provide the reader with an extensive understanding of the topic.

Habitat

A habitat is an ecological or environmental area that is inhabited by a particular species of animal, plant, or other type of organism. The term typically refers to the zone in which the organism lives and where it can find food, shelter, protection and mates for reproduction. It is the natural environment in which an organism lives, or the physical environment that surrounds a species population.

This coral reef in the Phoenix Islands Protected Area is a rich habitat for sea life.

A habitat is made up of physical factors such as soil, moisture, range of temperature, and light intensity as well as biotic factors such as the availability of food and the presence or absence of predators. Every organism has certain habitat needs for the conditions in which it will thrive, but some are tolerant of wide variations while others are very specific in their requirements. A habitat is not necessarily a geographical area, it can be the interior of a stem, a rotten log, a rock or a clump of moss, and for a parasitic organism it is the body of its host, part of the host's body such as the digestive tract, or a single cell within the host's body.

Habitat types include polar, temperate, subtropical and tropical. The terrestrial vegetation type may be forest, steppe, grassland, semi-arid or desert. Fresh water habitats include marshes, streams, rivers, lakes, ponds and estuaries, and marine habitats include salt marshes, the coast, the intertidal zone, reefs, bays, the open sea, the sea bed, deep water and submarine vents.

Few creatures make the ice shelves of Antarctica their habitat.

Ibex in alpine habitat.

Habitats change over time. This may be due to a violent event such as the eruption of a volcano, an earthquake, a tsunami, a wildfire or a change in oceanic currents; or the change may be more gradual over millennia with alterations in the climate, as ice sheets and glaciers advance and retreat, and as different weather patterns bring changes of precipitation and solar radiation. Other changes come as a direct result of human activities; deforestation, the ploughing of ancient grasslands, the diversion and damming of rivers, the draining of marshland and the dredging of the seabed. The introduction of alien species can have a devastating effect on native wildlife, through increased predation, through competition for resources or through the introduction of pests and diseases to which the native species have no immunity.

Definition and Etymology

The word "habitat" has been in use since about 1755 and derives from the Latin third-person singular present indicative of *habitāre*, to inhabit, from *habēre*, to have or to hold. Habitat can be defined as the natural environment of an organism, the place in which it is natural for it to live and grow. It is similar in meaning to a biotope, an area of uniform environmental conditions associated with a particular community of plants and animals.

Environmental Factors

The chief environmental factors affecting the distribution of living organisms are temperature, humidity, climate, soil type and light intensity, and the presence or absence of all the requirements that the organism needs to sustain it. Generally speaking, animal communities are reliant on specific types of plant communities.

Some plants and animals are generalists, and their habitat requirements are met in a wide range of locations. The small white butterfly (*Pieris rapae*) for example is found on all the continents of the world apart from Antarctica. Its larvae feed on a wide range of *Brassicas* and various other plant species, and it thrives in any open location with diverse plant associations. The large blue butterfly is much more specific in its requirements; it is found only in chalk grassland areas, its larvae feed on *Thymus* species and because of complex lifecycle requirements it inhabits only areas in which *Myrmica* ants live.

Disturbance is important in the creation of biodiverse habitats. In the absence of disturbance, a climax vegetation cover develops that prevents the establishment of other species. Wildflower meadows are sometimes created by conservationists but most of the flowering plants used are either annuals or biennials and disappear after a few years in the absence of patches of bare ground on which their seedlings can grow. Lightning strikes and toppled trees in tropical forests allow species richness to be maintained as pioneering species move in to fill the gaps created. Similarly coastal habitats can become dominated by kelp until the seabed is disturbed by a storm and the algae swept away, or shifting sediment exposes new areas for colonisation. Another cause of disturbance is when an area may be overwhelmed by an invasive introduced species which is not kept under control by natural enemies in its new habitat.

Types of Habitat

Rich rainforest habitat in Dominica.

Terrestrial habitat types include forests, grasslands, wetlands and deserts. Within these broad biomes are more specific habitats with varying climate types, temperature regimes, soils, altitudes and vegetation types. Many of these habitats grade into each other and each one has its own typical communities of plants and animals. A habitat may suit a particular species well, but its presence or absence at any particular location depends to some extent on chance, on its dispersal abilities and its efficiency as a coloniser.

Wetland habitats in Borneo.

Freshwater habitats include rivers, streams, lakes, ponds, marshes and bogs. Although some organisms are found across most of these habitats, the majority have more specific requirements. The water velocity, its temperature and oxygen saturation are important factors, but in river systems, there are fast and slow sections, pools, bayous and backwaters which provide a range of habitats. Similarly, aquatic plants can be floating, semi-submerged, submerged or grow in permanently or temporarily saturated soils besides bodies of water. Marginal plants provide important habitat for both invertebrates and vertebrates, and submerged plants provide oxygenation of the water, absorb nutrients and play a part in the reduction of pollution.

Marine habitats include brackish water, estuaries, bays, the open sea, the intertidal zone, the sea bed, reefs and deep water zones. Further variations include rock pools, sand banks, mudflats, brackish lagoons, sandy and pebbly beaches, and seagrass beds, all supporting their own flora and fauna. The benthic zone or seabed provides a home for both static organisms, anchored to the substrate, and for a large range of organisms crawling on or burrowing into the surface. Some creatures float among the waves on the surface of the water, or raft on floating debris, others swim at a range of depths, including organisms in the demersal zone close to the seabed, and myriads of organisms drift with the currents and form the plankton.

Desert scene in Egypt.

A desert is not the kind of habitat that favours the presence of amphibians, with their requirement for water to keep their skins moist and for the development of their young. Nevertheless, some frogs live in deserts, creating moist habitats underground and hibernating while conditions are adverse. Couch's spadefoot toad (*Scaphiopus couchii*) emerges from its burrow when a downpour occurs and lays its eggs in the transient pools that form; the tadpoles develop with great rapidity, sometimes in as little as nine days, undergo metamorphosis, and feed voraciously before digging a burrow of their own.

Other organisms cope with the drying up of their aqueous habitat in other ways. Vernal pools are ephemeral ponds that form in the rainy season and dry up afterwards. They have their specially-adapted characteristic flora, mainly consisting of annuals, the seeds of which survive the drought, but also some uniquely adapted perennials. Animals adapted to these extreme habitats also exist; fairy shrimps can lay "winter eggs" which are resistant to desiccation, sometimes being blown about with the dust, ending up in new depressions in the ground. These can survive in a dormant state for as long as fifteen years. Some killifish behave in a similar way; their eggs hatch and the juvenile fish grow with great rapidity when the conditions are right, but the whole population of fish may end up as eggs in diapause in the dried up mud that was once a pond.

Many animals and plants have taken up residence in urban environments. They tend to be adaptable generalists and use the town's features to make their homes. Rats and mice have followed man around the globe, pigeons, peregrines, sparrows, swallows and house martins use the buildings for nesting, bats use roof space for roosting, foxes visit the garbage bins and squirrels, coyotes, raccoons and skunks roam the streets. About 2,000 coyotes are thought to live in and around Chicago. A survey of dwelling houses in northern European cities in the twentieth century found about 175 species of invertebrate inside them, including 53 species of beetle, 21 flies, 13 butterflies and moths, 13 mites, 9 lice, 7 bees, 5 wasps, 5 cockroaches, 5 spiders, 4 ants and a number of other groups. In warmer climates, termites are serious pests in the urban habitat; 183 species are known to affect buildings and 83 species cause serious structural damage.

Microhabitats

A microhabitat is the small-scale physical requirements of a particular organism or population. Every habitat includes large numbers of microhabitats with subtly different exposure to light, humidity, temperature, air movement, and other factors. The lichens that grow on the north face of a boulder are different to those that grow on the south face, from those on the level top and those that grow on the ground nearby; the lichens growing in the grooves and on the raised surfaces are different from those growing on the veins of quartz. Lurking among these miniature "forests" are the microfauna, each species of invertebrate with its own specific habitat requirements.

There are numerous different microhabitats in a wood; coniferous forest, broad-leafed forest, open woodland, scattered trees, woodland verges, clearings and glades; tree trunk, branch, twig, bud, leaf, flower and fruit; rough bark, smooth bark, damaged bark, rotten wood, hollow, groove and hole; canopy, shrub layer, plant layer, leaf litter and soil; buttress root, stump, fallen log, stem base, grass tussock, fungus, fern and moss. The greater the structural diversity in the wood, the greater the number of microhabitats that will be present. A range of tree species with individual specimens of varying sizes and ages, and a range of features such as streams, level areas, slopes, tracks, clearings and felled areas will provide suitable conditions for an enormous number of biodiverse plants and animals. For example, in Britain it has been estimated that various types of rotting wood are home to over 1700 species of invertebrate.

For a parasitic organism, its habitat is the particular part of the outside or inside of its host on or in which it is adapted to live. The life cycle of some parasites involves several different host species, as well as free-living life stages, sometimes providing vastly different microhabitats. One such organism is the trematode (flatworm) *Microphallus turgidus*, present in brackish water marshes in the southeastern United States. Its first intermediate host is a snail and the second, a glass shrimp. The final host is the waterfowl or mammal that consumes the shrimp.

Extreme Habitats

Although the vast majority of life on Earth lives in mesophyllic (moderate) environments, a few organisms, most of them microbes, have managed to colonise extreme environments that are unsuitable for most higher life forms. There are bacteria, for example, living in Lake Whillans, half a mile below the ice of Antarctica; in the absence of sunlight, they must rely on organic material from elsewhere, perhaps decaying matter from glacier melt water or minerals from the underlying rock. Other bacteria can be found in abundance in the Mariana Trench, the deepest place in the ocean and on Earth; marine snow drifts down from the surface layers of the sea and accumulates in this undersea valley, providing nourishment for an extensive community of bacteria.

An Antarctic rock split apart to show an endolithic lifeform showing as a green layer a few millimetres thick.

Other microbes live in habitats lacking in oxygen, and are dependent on chemical reactions other than photosynthesis. Boreholes drilled 300 m (1,000 ft) into the rocky seabed have found microbial communities apparently based on the products of reactions between water and the constituents of rocks. These communities have been little studied, but may be an important part of the global carbon cycle. Rock in mines two miles deep also harbour microbes; these live on minute traces of hydrogen produced in slow oxidizing reactions inside the rock. These metabolic reactions allow life to exist in places with no oxygen or light, an environment that had previously been thought to be devoid of life.

The intertidal zone and the photic zone in the oceans are relatively familiar habitats. However the vast bulk of the ocean is unhospitable to air-breathing humans, with scuba divers limited to the upper 50 m (160 ft) or so. The lower limit for photosynthesis is 100 to 200 m (330 to 660 ft) and below that depth the prevailing conditions include total darkness, high pressure, little oxygen (in some places), scarce food resources and extreme cold. This habitat is very challenging to research, and as well as being little studied, it is vast, with 79% of the Earth's biosphere being at depths greater than 1,000 m (3,300 ft). With no plant life, the animals in this zone are either detritivores, reliant on food drifting down from surface layers, or they are predators, feeding on each other. Some organisms are pelagic, swimming or drifting in mid-ocean, while others are benthic, living on or near the seabed. Their growth rates and metabolisms tend to be slow, their eyes may be very large to detect what little illumination there is, or they may be blind and rely on other sensory inputs. A number of deep sea creatures are bioluminescent; this serves a variety of functions including predation, protection and social recognition. In general, the bodies of animals living at great depths are adapted to high pressure environments by having pressure-resistant biomolecules and small organic molecules present in their cells know as piezolytes, which give the proteins the flexibility they need. There are also unsaturated fats in their membranes which prevent them from solidifying at low temperatures.

Dense mass of white crabs at a hydrothermal vent, with stalked barnacles on right.

Hydrothermal vents were first discovered in the ocean depths in 1977. They result from seawater becoming heated after seeping through cracks to places where hot magma is close to the seabed. The under-water hot springs may gush forth at temperatures of over 340 °C (640 °F) and support unique communities of organisms in their immediate vicinity. The basis for this teeming life is chemosynthesis, a process by which microbes convert such substances as hydrogen sulfide or ammonia into organic molecules. These bacteria and Archaea are the primary producers in these ecosystems and support a diverse array of life. About 350 species of organism, dominated by molluscs, polychaete worms and crustaceans, had been discovered around hydrothermal vents by the end of the twentieth century, most of them being new to science and endemic to these habitats.

Besides providing locomotion opportunities for winged animals and a conduit for the dispersal of pollen grains, spores and seeds, the atmosphere can be considered to be a habitat in its own right. There are metabolically active microbes present that actively reproduce and spend their whole existence airborne, with hundreds of thousands of individual organisms estimated to be present in a cubic metre of air. The airborne microbial community may be as diverse as that found in soil or other terrestrial environments, however these organisms are not evenly distributed, their densities varying spatially with altitude and environmental conditions. Aerobiology has been little studied, but there is evidence of nitrogen fixation in clouds, and less clear evidence of carbon cycling, both facilitated by microbial activity.

There are other examples of extreme habitats where specially adapted lifeforms exist; tar pits teeming with microbial life; naturally occurring crude oil pools inhabited by the larvae of the petroleum fly; hot springs where the temperature may be as high as 71 °C (160 °F) and cyanobacteria create microbial mats; cold seeps where the methane and hydrogen sulfide issue from the ocean floor and support microbes and higher animals such as mussels which form symbiotic associations with these anaerobic organisms; salt pans harbour salt-tolerant microorganisms and also *Wallemia ichthyophaga*, a basidomycotous fungus; ice sheets in Antarctica which support fungi *Thelebolus* spp., and snowfields on which algae grow.

Habitat Change

Whether from natural processes or the activities of man, landscapes and their associated habitats change over time. There are the slow geomorphological changes associated with the geologic processes that cause tectonic uplift and subsidence, and the more rapid changes associated with earthquakes, landslides, storms, flooding, wildfires, coastal erosion, deforestation and changes in land use. Then there are the changes in habitats brought on by alterations in farming practices, tourism, pollution, fragmentation and climate change.

Twenty five years after the devastating eruption at Mount St. Helens, United States, pioneer species have moved in.

Loss of habitat is the single greatest threat to any species. If an island on which an endemic organism lives becomes uninhabitable for some reason, the species will become extinct. Any type of habitat surrounded by a different habitat is in a similar situation to an island. If a forest is divided into parts by logging, with strips of cleared land separating woodland blocks, and the distances between the remaining fragments exceeds the distance an individual animal is able to travel, that species becomes especially vulnerable. Small populations generally lack genetic diversity and may be threatened by increased predation, increased competition, disease and unexpected catastrophe. At the edge of each forest fragment, increased light encourages secondary growth of fast-growing species and old growth trees are more vulnerable to logging as access is improved. The birds that nest in their crevices, the epiphytes that hang from their branches and the invertebrates in the leaf litter are all adversely affected and biodiversity is reduced. Habitat fragmentation can be ameliorated to some extent by the provision of wildlife corridors connecting the fragments. These can be a river, ditch, strip of trees, hedgerow or even an underpass to a highway. Without the corridors, seeds cannot disperse and animals, especially small ones, cannot travel through the hostile territory, putting populations at greater risk of local extinction.

Habitat disturbance can have long-lasting effects on the environment. *Bromus tectorum* is a vigorous grass from Europe which has been introduced to the United States where it has become invasive. It is highly adapted to fire, producing large amounts of flammable detritus and increasing the frequency and intensity of wildfires. In areas where it has become established, it has altered the local fire regimen to such an extant that native plants cannot survive the frequent fires, allowing it to become even more dominant. A marine example is when sea urchin populations "explode" in coastal waters and destroy all the macroalgae present. What was previously a kelp forest becomes an urchin barren that may last for years and this can have a profound effect on the food chain.

Removal of the sea urchins, by disease for example, can result in the seaweed returning, with an over-abundance of fast-growing kelp.

Habitat Protection

The protection of habitats is a necessary step in the maintenance of biodiversity because if habitat destruction occurs, the animals and plants reliant on that habitat suffer. Many countries have enacted legislation to protect their wildlife. This may take the form of the setting up of national parks, forest reserves and wildlife reserves, or it may restrict the activities of humans with the objective of benefiting wildlife. The laws may be designed to protect a particular species or group of species, or the legislation may prohibit such activities as the collecting of bird eggs, the hunting of animals or the removal of plants. A general law on the protection of habitats may be more difficult to implement than a site specific requirement. A concept introduced in the Unites States in 1973 involves protecting the critical habitat of endangered species, and a similar concept has been incorporated into some Australian legislation.

International treaties may be necessary for such objectives as the setting up of marine reserves. Another international agreement, the Convention on the Conservation of Migratory Species of Wild Animals, protects animals that migrate across the globe and need protection in more than one country. However, the protection of habitats needs to take into account the needs of the local residents for food, fuel and other resources. Even where legislation protects the environment, a lack of enforcement often prevents effective protection. Faced with food shortage, a farmer is likely to plough up a level patch of ground despite it being the last suitable habitat for an endangered species such as the San Quintin kangaroo rat, and even kill the animal as a pest. In this regard, it is desirable to educate the community on the uniqueness of their flora and fauna and the benefits of ecotourism.

Monotypic Habitat

A monotypic habitat is one in which a single species of animal or plant is so dominant as to virtually exclude all other species. An example would be sugarcane; this is planted, burnt and harvested, with herbicides killing weeds and pesticides controlling invertebrates. The monotypic habitat occurs in botanical and zoological contexts, and is a component of conservation biology. In restoration ecology of native plant communities or habitats, some invasive species create monotypic stands that replace and/or prevent other species, especially indigenous ones, from growing there. A dominant colonization can occur from retardant chemicals exuded, nutrient monopolization, or from lack of natural controls such as herbivores or climate, that keep them in balance with their native habitats. The yellow starthistle, *Centaurea solstitialis*, is a botanical monotypic-habitat example of this, currently dominating over 15,000,000 acres (61,000 km^2) in California alone. The non-native freshwater zebra mussel, *Dreissena polymorpha*, that colonizes areas of the Great Lakes and the Mississippi River watershed, is a zoological monotypic-habitat example; the predators that control it in its home-range in Russia are absent and it proliferates abundantly. Even though its name may seem to imply simplicity as compared with polytypic habitats, the monotypic habitat can be complex. Aquatic habitats, such as exotic *Hydrilla* beds, support a similarly rich fauna of macroinvertebrates to a more varied habitat, but the creatures present may differ between the two, affecting small fish and other animals higher up the food chain.

Biodiversity

Biodiversity, a contraction of "biological diversity," generally refers to the variety and variability of life on Earth. One of the most widely used definitions defines it in terms of the variability within species, between species and between ecosystems. It is a measure of the variety of organisms present in different ecosystems. This can refer to genetic variation, ecosystem variation, or species variation (number of species) within an area, biome, or planet. Terrestrial biodiversity tends to be greater near the equator, which seems to be the result of the warm climate and high primary productivity. Biodiversity is not distributed evenly on Earth. It is richest in the tropics. Marine biodiversity tends to be highest along coasts in the Western Pacific, where sea surface temperature is highest and in the mid-latitudinal band in all oceans. There are latitudinal gradients in species diversity. Biodiversity generally tends to cluster in hotspots, and has been increasing through time, but will be likely to slow in the future.

The number and variety of plants, animals and other organisms that exist is known as biodiversity. It is an essential component of nature and it ensures the survival of human species by providing food, fuel, shelter, medicines and other resources to mankind. The richness of biodiversity depends on the climatic conditions and area of the region. All species of plants taken together are known as flora and about 70,000 species of plants are known to date. All species of animals taken together are known as fauna which includes birds, mammals, fish, reptiles, insects, crustaceans, molluscs, etc.

Rapid environmental changes typically cause mass extinctions. More than 99 percent of all species, amounting to over five billion species, that ever lived on Earth are estimated to be extinct. Estimates on the number of Earth's current species range from 10 million to 14 million, of which about 1.2 million have been documented and over 86 percent have not yet been described. More recently, in May 2016, scientists reported that 1 trillion species are estimated to be on Earth currently with only one-thousandth of one percent described. The total amount of related DNA base pairs on Earth is estimated at 5.0×10^{37} and weighs 50 billion tonnes. In comparison, the total mass of the biosphere has been estimated to be as much as 4 TtC (trillion tons of carbon). In July 2016, scientists reported identifying a set of 355 genes from the Last Universal Common Ancestor (LUCA) of all organisms living on Earth.

The age of the Earth is about 4.54 billion years old. The earliest undisputed evidence of life on Earth dates at least from 3.5 billion years ago, during the Eoarchean Era after a geological crust started to solidify following the earlier molten Hadean Eon. There are microbial mat fossils found in 3.48 billion-year-old sandstone discovered in Western Australia. Other early physical evidence of a biogenic substance is graphite in 3.7 billion-year-old meta-sedimentary rocks discovered in Western Greenland. More recently, in 2015, "remains of biotic life" were found in 4.1 billion-year-old rocks in Western Australia. According to one of the researchers, "If life arose relatively quickly on Earth .. then it could be common in the universe."

Since life began on Earth, five major mass extinctions and several minor events have led to large and sudden drops in biodiversity. The Phanerozoic eon (the last 540 million years) marked a rapid growth in biodiversity via the Cambrian explosion—a period during which the majority of multicellular phyla first appeared. The next 400 million years included repeated, massive biodiversity losses classified as mass extinction events. In the Carboniferous, rainforest collapse led to a great loss of plant and animal life. The Permian–Triassic extinction event, 251 million years ago, was the worst; vertebrate recovery took 30 million years. The most recent, the Cretaceous–Paleogene extinction event, occurred 65 million years ago and has often attracted more attention than others because it resulted in the extinction of the dinosaurs.

The period since the emergence of humans has displayed an ongoing biodiversity reduction and an accompanying loss of genetic diversity. Named the Holocene extinction, the reduction is caused primarily by human impacts, particularly habitat destruction. Conversely, biodiversity impacts human health in a number of ways, both positively and negatively.

The United Nations designated 2011–2020 as the United Nations Decade on Biodiversity.

Etymology

The term biological diversity was used first by wildlife scientist and conservationist Raymond F. Dasmann in the year 1968 lay book *A Different Kind of Country* advocating conservation. The term was widely adopted only after more than a decade, when in the 1980s it came into common usage in science and environmental policy. Thomas Lovejoy, in the foreword to the book *Conservation Biology*, introduced the term to the scientific community. Until then the term "natural diversity" was common, introduced by The Science Division of The Nature Conservancy in an important 1975 study, "The Preservation of Natural Diversity." By the early 1980s TNC's Science program and its head, Robert E. Jenkins, Lovejoy and other leading conservation scientists at the time in America advocated the use of the term "biological diversity".

The term's contracted form *biodiversity* may have been coined by W.G. Rosen in 1985 while planning the 1986 *National Forum on Biological Diversity* organized by the National Research Council (NRC). It first appeared in a publication in 1988 when sociobiologist E. O. Wilson used it as the title of the proceedings of that forum.

Since this period the term has achieved widespread use among biologists, environmentalists, political leaders and concerned citizens.

A similar term in the United States is *"natural heritage."* It pre-dates the others and is more accepted by the wider audience interested in conservation. Broader than biodiversity, it includes geology and landforms.

Definitions

"Biodiversity" is most commonly used to replace the more clearly defined and long established terms, species diversity and species richness. Biologists most often define biodiversity as the "totality of genes, species and ecosystems of a region". An advantage of this definition is that it seems to describe most circumstances and presents a unified view of the traditional types of biological variety previously identified:

- Taxonomic diversity (usually measured at the species diversity level).
- Ecological diversity often viewed from the perspective of ecosystem diversity.
- Morphological diversity which stems from genetic diversity.

- Functional diversity which is a measure of the number of functionally disparate species within a population (e.g. different feeding mechanism, different motility, predator vs prey, etc.)

A sampling of fungi collected during summer 2008 in Northern Saskatchewan mixed woods, near LaRonge is an example regarding the species diversity of fungi. In this photo, there are also leaf lichens and mosses.

In 2003 Professor Anthony Campbell at Cardiff University, UK and the Darwin Centre, Pembrokeshire, defined a fourth level: Molecular Diversity.

This multilevel construct is consistent with Datman and Lovejoy. An explicit definition consistent with this interpretation was first given in a paper by Bruce A. Wilcox commissioned by the International Union for the Conservation of Nature and Natural Resources (IUCN) for the 1982 World National Parks Conference. Wilcox's definition was "Biological diversity is the variety of life forms...at all levels of biological systems (i.e., molecular, organismic, population, species and ecosystem)...". The 1992 United Nations Earth Summit defined "biological diversity" as "the variability among living organisms from all sources, including, 'inter alia', terrestrial, marine and other aquatic ecosystems and the ecological complexes of which they are part: this includes diversity within species, between species and of ecosystems". This definition is used in the United Nations Convention on Biological Diversity.

One textbook's definition is "variation of life at all levels of biological organization".

Genetically biodiversity can be defined as the diversity of alleles, genes and organisms. They study processes such as mutation and gene transfer that drive evolution.

Measuring diversity at one level in a group of organisms may not precisely correspond to diversity at other levels. However, tetrapod (terrestrial vertebrates) taxonomic and ecological diversity shows a very close correlation.

Distribution

Biodiversity is not evenly distributed, rather it varies greatly across the globe as well as within regions. Among other factors, the diversity of all living things (biota) depends on temperature, precipitation, altitude, soils, geography and the presence of other species. The study of the spatial distribution of organisms, species and ecosystems, is the science of biogeography.

Diversity consistently measures higher in the tropics and in other localized regions such as the Cape Floristic Region and lower in polar regions generally. Rain forests that have had wet climates for a long time, such as Yasuní National Park in Ecuador, have particularly high biodiversity.

Terrestrial biodiversity is thought to be up to 25 times greater than ocean biodiversity. A recently discovered method put the total number of species on Earth at 8.7 million, of which 2.1 million were estimated to live in the ocean. However, this estimate seems to under-represent the diversity of microorganisms.

Latitudinal Gradients

Generally, there is an increase in biodiversity from the poles to the tropics. Thus localities at lower latitudes have more species than localities at higher latitudes. This is often referred to as the latitudinal gradient in species diversity. Several ecological mechanisms may contribute to the gradient, but the ultimate factor behind many of them is the greater mean temperature at the equator compared to that of the poles.

Even though terrestrial biodiversity declines from the equator to the poles, some studies claim that this characteristic is unverified in aquatic ecosystems, especially in marine ecosystems. The latitudinal distribution of parasites does not appear to follow this rule.

Hotspots

A biodiversity hotspot is a region with a high level of endemic species that has experienced great habitat loss. The term hotspot was introduced in 1988 by Norman Myers. While hotspots are spread all over the world, the majority are forest areas and most are located in the tropics.

Brazil's Atlantic Forest is considered one such hotspot, containing roughly 20,000 plant species, 1,350 vertebrates and millions of insects, about half of which occur nowhere else. The island of Madagascar and India are also particularly notable. Colombia is characterized by high biodiversity, with the highest rate of species by area unit worldwide and it has the largest number of endemics (species that are not found naturally anywhere else) of any country. About 10% of the species of the Earth can be found in Colombia, including over 1,900 species of bird, more than in Europe and North America combined, Colombia has 10% of the world's mammals species, 14% of the amphibian species and 18% of the bird species of the world. Madagascar dry deciduous forests and lowland rainforests possess a high ratio of endemism. Since the island separated from mainland Africa 66 million years ago, many species and ecosystems have evolved independently. Indonesia's 17,000 islands cover 735,355 square miles (1,904,560 km^2) and contain 10% of the world's flowering plants, 12% of mammals and 17% of reptiles, amphibians and birds—along with nearly 240 million people. Many regions of high biodiversity and/or endemism arise from specialized habitats which require unusual adaptations, for example, alpine environments in high mountains, or Northern European peat bogs.

Accurately measuring differences in biodiversity can be difficult. Selection bias amongst researchers may contribute to biased empirical research for modern estimates of biodiversity. In 1768,

Rev. Gilbert White succinctly observed of his Selborne, Hampshire *"all nature is so full, that that district produces the most variety which is the most examined."*

Evolution and History

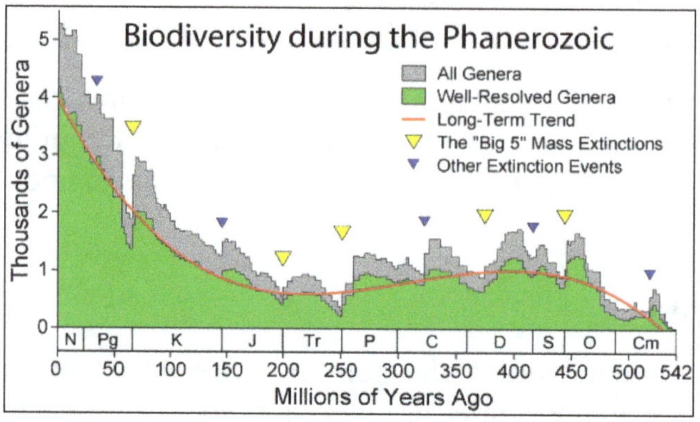

Apparent marine fossil diversity during the Phanerozoic.

Biodiversity is the result of 3.5 billion years of evolution. The origin of life has not been definitely established by science, however some evidence suggests that life may already have been well-established only a few hundred million years after the formation of the Earth. Until approximately 600 million years ago, all life consisted of archaea, bacteria, protozoans and similar single-celled organisms.

The history of biodiversity during the Phanerozoic (the last 540 million years), starts with rapid growth during the Cambrian explosion—a period during which nearly every phylum of multicellular organisms first appeared. Over the next 400 million years or so, invertebrate diversity showed little overall trend and vertebrate diversity shows an overall exponential trend. This dramatic rise in diversity was marked by periodic, massive losses of diversity classified as mass extinction events. A significant loss occurred when rainforests collapsed in the carboniferous. The worst was the Permian-Triassic extinction event, 251 million years ago. Vertebrates took 30 million years to recover from this event.

The fossil record suggests that the last few million years featured the greatest biodiversity in history. However, not all scientists support this view, since there is uncertainty as to how strongly the fossil record is biased by the greater availability and preservation of recent geologic sections. Some scientists believe that corrected for sampling artifacts, modern biodiversity may not be much different from biodiversity 300 million years ago., whereas others consider the fossil record reasonably reflective of the diversification of life. Estimates of the present global macroscopic species diversity vary from 2 million to 100 million, with a best estimate of somewhere near 9 million, the vast majority arthropods. Diversity appears to increase continually in the absence of natural selection.

Evolutionary Diversification

The existence of a "global carrying capacity", limiting the amount of life that can live at once, is debated, as is the question of whether such a limit would also cap the number of species. While records of life in the sea shows a logistic pattern of growth, life on land (insects, plants and tetrapods) shows an exponential rise in diversity. As one author states, "Tetrapods have not yet invaded 64 per cent of potentially habitable modes and it could be that without human influence the ecological and taxonomic diversity of tetrapods would continue to increase in an exponential fashion until most or all of the available ecospace is filled."

On the other hand, changes through the Phanerozoic correlate much better with the hyperbolic model (widely used in population biology, demography and macrosociology, as well as fossil biodiversity) than with exponential and logistic models. The latter models imply that changes in diversity are guided by a first-order positive feedback (more ancestors, more descendants) and/ or a negative feedback arising from resource limitation. Hyperbolic model implies a second-order positive feedback. The hyperbolic pattern of the world population growth arises from a second-order positive feedback between the population size and the rate of technological growth. The hyperbolic character of biodiversity growth can be similarly accounted for by a feedback between diversity and community structure complexity. The similarity between the curves of biodiversity and human population probably comes from the fact that both are derived from the interference of the hyperbolic trend with cyclical and stochastic dynamics.

Most biologists agree however that the period since human emergence is part of a new mass extinction, named the Holocene extinction event, caused primarily by the impact humans are having on the environment. It has been argued that the present rate of extinction is sufficient to eliminate most species on the planet Earth within 100 years.

New species are regularly discovered (on average between 5–10,000 new species each year, most of them insects) and many, though discovered, are not yet classified (estimates are that nearly 90% of all arthropods are not yet classified). Most of the terrestrial diversity is found in tropical forests and in general, land has more species than the ocean; some 8.7 million species may exists on Earth, of which some 2.1 million live in the ocean.

Biodiversity and Ecosystem Services

Summer field in Belgium (Hamois). The blue flowers are *Centaurea cyanus* and the red are *Papaver rhoeas*.

The Balance of Evidence

"Ecosystem services are the suite of benefits that ecosystems provide to humanity."

These services come in three flavors:

1. Provisioning services which involve the production of renewable resources (e.g.: food, wood, fresh water).
2. Regulating services which are those that lessen environmental change (e.g.: climate regulation, pest/disease control).
3. Cultural services represent human value and enjoyment (e.g.: landscape aesthetics, cultural heritage, outdoor recreation and spiritual significance).

There have been many claims about biodiversity's effect on these ecosystem services, especially provisioning and regulating services. After an exhaustive survey through peer-reviewed literature to evaluate 36 different claims about biodiversity's effect on ecosystem services, 14 of those claims have been validated, 6 demonstrate mixed support or are unsupported, 3 are incorrect and 13 lack enough evidence to draw definitive conclusions.

Services Enhanced by Biodiversity

Provisioning Services

- Greater species diversity of plants increases fodder yield (synthesis of 271 experimental studies).
- Greater genetic diversity of plants (i.e.: diversity within a single species) increases overall crop yield (synthesis of 575 experimental studies). Although another review of 100 experimental studies reports mixed evidence.
- Greater species diversity of trees increases overall wood production (Synthesis of 53 experimental studies). However, there is not enough data to draw a conclusion about the effect of tree trait diversity on wood production.

Regulating Services

- Greater species diversity of fish increases the stability of fisheries yield (Synthesis of 8 observational studies).
- Greater species diversity of natural pest enemies decreases herbivorous pest populations (Data from two separate reviews; Synthesis of 266 experimental and observational studies; Synthesis of 18 observational studies. Although another review of 38 experimental studies found mixed support for this claim, suggesting that in cases where mutual intraguild predation occurs, a single predatory species is often more effective.
- Greater species diversity of plants decreases disease prevalence on plants (Synthesis of 107 experimental studies).
- Greater species diversity of plants increases resistance to plant invasion (Data from two separate reviews; Synthesis of 105 experimental studies; Synthesis of 15 experimental studies).

- Greater species diversity of plants increases carbon sequestration, but note that this finding only relates to actual uptake of carbon dioxide and not long term storage, Synthesis of 479 experimental studies).
- Greater species diversity of plants increases soil nutrient remineralization (Synthesis of 103 experimental studies).
- Greater species diversity of plants increases soil organic matter (Synthesis of 85 experimental studies).

Services with Mixed Evidence

Provisioning Services

- None to date.

Regulating Services

- Greater species diversity of plants may or may not decrease herbivorous pest populations. Data from two separate reviews suggest that greater diversity decreases pest populations (Synthesis of 40 observational studies; Synthesis of 100 experimental studies). One review found mixed evidence (Synthesis of 287 experimental studies), while another found contrary evidence (Synthesis of 100 experimental studies).
- Greater species diversity of animals may or may not decrease disease prevalence on those animals (Synthesis of 45 experimental and observational studies), although a 2013 study offers more support showing that biodiversity may in fact enhance disease resistance within animal communities, at least in amphibian frog ponds. Many more studies must be published in support of diversity to sway the balance of evidence will be such that we can draw a general rule on this service.
- Greater species and trait diversity of plants may or may not increase long term carbon storage (Synthesis of 33 observational studies).
- Greater pollinator diversity may or may not increase pollination (Synthesis of 7 observational studies), but a publication from March 2013 suggests that increased native pollinator diversity enhances pollen deposition (although not necessarily fruit set as the authors would have you believe, for details explore their lengthy supplementary material).

Services for Which Biodiversity is a Hindrance

Provisioning Services

- Greater species diversity of plants reduces primary production (Synthesis of 7 experimental studies).

Regulating Services

- Greater genetic and species diversity of a number of organisms reduces freshwater purification (Synthesis of 8 experimental studies, although an attempt by the authors to investigate the effect of detritivore diversity on freshwater purification was unsuccessful due to a lack of available evidence (only 1 observational study was found).

Provisioning Services

- Effect of species diversity of plants on biofuel yield (In a survey of the literature, the investigators only found 3 studies).
- Effect of species diversity of fish on fishery yield (In a survey of the literature, the investigators only found 4 experimental studies and 1 observational study).

Regulating Services

- Effect of species diversity on the stability of biofuel yield (In a survey of the literature, the investigators did not find any studies).
- Effect of species diversity of plants on the stability of fodder yield (In a survey of the literature, the investigators only found 2 studies).
- Effect of species diversity of plants on the stability of crop yield (In a survey of the literature, the investigators only found 1 study).
- Effect of genetic diversity of plants on the stability of crop yield (In a survey of the literature, the investigators only found 2 studies).
- Effect of diversity on the stability of wood production (In a survey of the literature, the investigators could not find any studies).
- Effect of species diversity of multiple taxa on erosion control (In a survey of the literature, the investigators could not find any studies – they did however find studies on the effect of species diversity and root biomass).
- Effect of diversity on flood regulation (In a survey of the literature, the investigators could not find any studies).
- Effect of species and trait diversity of plants on soil moisture (In a survey of the literature, the investigators only found 2 studies).

Other sources have reported somewhat conflicting results and in 1997 Robert Costanza and colleagues reported the estimated global value of ecosystem services (not captured in traditional markets) at an average of $33 trillion annually.

Since the stone age, species loss has accelerated above the average basal rate, driven by human activity. Estimates of species losses are at a rate 100-10,000 times as fast as is typical in the fossil record. Biodiversity also affords many non-material benefits including spiritual and aesthetic values, knowledge systems and education.

Biodiversity and Agriculture

Agricultural diversity can be divided into two categories: intraspecific diversity, which includes the genetic variety within a single species, like the potato (*Solanum tuberosum*) that is composed of many different forms and types (e.g.: in the U.S. we might compare russet potatoes with new potatoes or purple potatoes, all different, but all part of the same species, *S. tuberosum*).

Amazon Rainforest in South America.

The other category of agricultural diversity is called interspecific diversity and refers to the number and types of different species. Thinking about this diversity we might note that many small vegetable farmers grow many different crops like potatoes and also carrots, peppers, lettuce etc.

Agricultural diversity can also be divided by whether it is 'planned' diversity or 'associated' diversity. This is a functional classification that we impose and not an intrinsic feature of life or diversity. Planned diversity includes the crops which a farmer has encouraged, planted or raised (e.g.: crops, covers, symbionts and livestock, among others), which can be contrasted with the associated diversity that arrives among the crops, uninvited (e.g.: herbivores, weed species and pathogens, among others).

The control of associated biodiversity is one of the great agricultural challenges that farmers face. On monoculture farms, the approach is generally to eradicate associated diversity using a suite of biologically destructive pesticides, mechanized tools and transgenic engineering techniques, then to rotate crops. Although some polyculture farmers use the same techniques, they also employ integrated pest management strategies as well as strategies that are more labor-intensive, but generally less dependent on capital, biotechnology and energy.

Interspecific crop diversity is, in part, responsible for offering variety in what we eat. Intraspecific diversity, the variety of alleles within a single species, also offers us choice in our diets. If a crop fails in a monoculture, we rely on agricultural diversity to replant the land with something new. If a wheat crop is destroyed by a pest we may plant a hardier variety of wheat the next year, relying on intraspecific diversity. We may forgo wheat production in that area and plant a different species altogether, relying on interspecific diversity. Even an agricultural society which primarily grows monocultures, relies on biodiversity at some point:

- The Irish potato blight of 1846 was a major factor in the deaths of one million people and the emigration of about two million. It was the result of planting only two potato varieties, both vulnerable to the blight, *Phytophthora infestans*, which arrived in 1845.
- When rice grassy stunt virus struck rice fields from Indonesia to India in the 1970s, 6,273 varieties were tested for resistance. Only one was resistant, an Indian variety and known to science only since 1966. This variety formed a hybrid with other varieties and is now widely grown.
- Coffee rust attacked coffee plantations in Sri Lanka, Brazil and Central America in 1970. A resistant variety was found in Ethiopia. The diseases are themselves a form of biodiversity.

Monoculture was a contributing factor to several agricultural disasters, including the European wine industry collapse in the late 19th century and the US southern corn leaf blight epidemic of 1970.

Although about 80 percent of humans' food supply comes from just 20 kinds of plants, humans use at least 40,000 species. Many people depend on these species for food, shelter and clothing. Earth's surviving biodiversity provides resources for increasing the range of food and other products suitable for human use, although the present extinction rate shrinks that potential.

Biodiversity and Human Health

Biodiversity's relevance to human health is becoming an international political issue, as scientific evidence builds on the global health implications of biodiversity loss. This issue is closely linked with the issue of climate change, as many of the anticipated health risks of climate change are associated with changes in biodiversity (e.g. changes in populations and distribution of disease vectors, scarcity of fresh water, impacts on agricultural biodiversity and food resources etc.) This is because the species most likely to disappear are those that buffer against infectious disease transmission, while surviving species tend to be the ones that increase disease transmission, such as that of West Nile Virus, Lyme disease and Hantavirus, according to a study done co-authored by Felicia Keesing, an ecologist at Bard College and Drew Harvell, associate director for Environment of the Atkinson Center for a Sustainable Future (ACSF) at Cornell University.

The diverse forest canopy on Barro Colorado Island, Panama, yielded this display of different fruit.

The growing demand and lack of drinkable water on the planet presents an additional challenge to the future of human health. Partly, the problem lies in the success of water suppliers to increase supplies and failure of groups promoting preservation of water resources. While the distribution of clean water increases, in some parts of the world it remains unequal. According to *2008 World Population Data Sheet*, only 62% of least developed countries are able to access clean water.

Some of the health issues influenced by biodiversity include dietary health and nutrition security, infectious disease, medical science and medicinal resources, social and psychological health. Biodiversity is also known to have an important role in reducing disaster risk and in post-disaster relief and recovery efforts.

Biodiversity provides critical support for drug discovery and the availability of medicinal resources. A significant proportion of drugs are derived, directly or indirectly, from biological sources: at least 50% of the pharmaceutical compounds on the US market are derived from plants, animals and micro-organisms, while about 80% of the world population depends on medicines from nature (used in either modern or traditional medical practice) for primary healthcare. Only a tiny fraction of wild species has been investigated for medical potential. Biodiversity has been critical to advances throughout the field of bionics. Evidence from market analysis and biodiversity science indicates that the decline in output from the pharmaceutical sector since the mid-1980s can be attributed to a move away from natural product exploration ("bioprospecting") in favor of genomics and synthetic chemistry, indeed claims about the value of undiscovered pharmaceuticals may not provide enough incentive for companies in free markets to search for them because of the high cost of development; meanwhile, natural products have a long history of supporting significant economic and health innovation. Marine ecosystems are particularly important, although inappropriate bioprospecting can increase biodiversity loss, as well as violating the laws of the communities and states from which the resources are taken.

Biodiversity, Business and Industry

Agriculture production, pictured is a tractor and a chaser bin.

Many industrial materials derive directly from biological sources. These include building materials, fibers, dyes, rubber and oil. Biodiversity is also important to the security of resources such as water, timber, paper, fiber and food. As a result, biodiversity loss is a significant risk factor in business development and a threat to long term economic sustainability.

Biodiversity, Leisure, Cultural and Aesthetic Value

Biodiversity enriches leisure activities such as hiking, birdwatching or natural history study. Biodiversity inspires musicians, painters, sculptors, writers and other artists. Many cultures view themselves as an integral part of the natural world which requires them to respect other living organisms.

Popular activities such as gardening, fishkeeping and specimen collecting strongly depend on biodiversity. The number of species involved in such pursuits is in the tens of thousands, though the majority do not enter commerce.

The relationships between the original natural areas of these often exotic animals and plants and commercial collectors, suppliers, breeders, propagators and those who promote their understanding and enjoyment are complex and poorly understood. The general public responds well to exposure to rare and unusual organisms, reflecting their inherent value.

Philosophically it could be argued that biodiversity has intrinsic aesthetic and spiritual value to mankind *in and of itself*. This idea can be used as a counterweight to the notion that tropical forests and other ecological realms are only worthy of conservation because of the services they provide.

Biodiversity and Ecological Services

Biodiversity supports many ecosystem services:

Eagle Creek, Oregon hiking.

"There is now unequivocal evidence that biodiversity loss reduces the efficiency by which ecological communities capture biologically essential resources, produce biomass, decompose and recycle biologically essential nutrients... There is mounting evidence that biodiversity increases the stability of ecosystem functions through time... Diverse communities are more productive because they contain key species that have a large influence on productivity and differences in functional traits among organisms increase total resource capture... The impacts of diversity loss on ecological processes might be sufficiently large to rival the impacts of many other global drivers of environmental change... Maintaining multiple ecosystem processes at multiple places and times requires higher levels of biodiversity than does a single process at a single place and time."

It plays a part in regulating the chemistry of our atmosphere and water supply. Biodiversity is directly involved in water purification, recycling nutrients and providing fertile soils. Experiments with controlled environments have shown that humans cannot easily build ecosystems to support human needs; for example insect pollination cannot be mimicked, and that activity alone represented between $2.1-14.6 billions in 2003.

Number of Species

Species	Earth			Ocean		
	Catalogued	Predicted	±SE	Catalogued	Predicted	±SE
Eukaryotes						
Animalia	953,434	7,770,000	958,000	171,082	2,150,000	145,000
Chromista	13,033	27,500	30,500	4,859	7,400	9,640
Fungi	43,271	611,000	297,000	1,097	5,320	11,100
Plantae	215,644	298,000	8,200	8,600	16,600	9,130
Protozoa	8,118	36,400	6,690	8,118	36,400	6,690
Total	1,233,500	8,740,000	1,300,000	193,756	2,210,000	182,000
Prokaryotes						
Archaea	502	455	160	1	1	0
Bacteria	10,358	9,680	3,470	652	1,320	436
Total	10,860	10,100	3,630	653	1,320	436
Grand Total	**1,244,360**	**8,750,000**	**1,300,000**	**194,409**	**2,210,000**	**182,000**

Predictions for prokaryotes represent a lower bound because they do not consider undescribed higher taxa. For protozoa, the ocean database was substantially more complete than the database for the entire Earth so we only used the former to estimate the total number of species in this taxon. All predictions were rounded to three significant digits.

Discovered and predicted total number of species on land and in the oceans.

According to Mora and colleagues, the total number of terrestrial species is estimated to be around 8.7 million while the number of oceanic species is much lower, estimated at 2.2 million. The authors note that these estimates are strongest for eukaryotic organisms and likely represent the lower bound of prokaryote diversity. Other estimates include:

- 220,000 vascular plants, estimated using the species-area relation method.
- 0.7-1 million marine species.
- 10–30 million insects; (of some 0.9 million we know today).
- 5–10 million bacteria.
- 1.5-3 million fungi, estimates based on data from the tropics, long-term non-tropical sites and molecular studies that have revealed cryptic speciation. Some 0.075 million species of fungi had been documented by 2001).
- 1 million mites.
- The number of microbial species is not reliably known, but the Global Ocean Sampling Expedition dramatically increased the estimates of genetic diversity by identifying an enormous number of new genes from near-surface plankton samples at various marine locations, initially over the 2004-2006 period. The findings may eventually cause a significant change in the way science defines species and other taxonomic categories.

Since the rate of extinction has increased, many extant species may become extinct before they are described.

Species Loss Rates

No longer do we have to justify the existence of humid tropical forests on the feeble grounds that they might carry plants with drugs that cure human disease. *Gaia theory* forces us to see that they offer much more than this. Through their capacity to evapotranspirate vast volumes of water vapor, they serve to keep the planet cool by wearing a sunshade of white reflecting cloud. Their replacement by cropland could precipitate a disaster that is global in scale.

During the last century, decreases in biodiversity have been increasingly observed. In 2007, German Federal Environment Minister Sigmar Gabriel cited estimates that up to 30% of all species will be extinct by 2050. Of these, about one eighth of known plant species are threatened with extinction. Estimates reach as high as 140,000 species per year (based on Species-area theory). This figure indicates unsustainable ecological practices, because few species emerge each year. Almost all scientists acknowledge that the rate of species loss is greater now than at any time in human history, with extinctions occurring at rates hundreds of times higher than background extinction rates. As of 2012, some studies suggest that 25% of all mammal species could be extinct in 20 years.

In absolute terms, the planet has lost 52% of its biodiversity since 1970 according to a 2014 study by the World Wildlife Fund. The Living Planet Report 2014 claims that "the number of mammals, birds, reptiles, amphibians and fish across the globe is, on average, about half the size it was 40 years ago". Of that number, 39% accounts for the terrestrial wildlife gone, 39% for the marine wildlife gone and 76% for the freshwater wildlife gone. Biodiversity took the biggest hit in Latin America, plummeting 83 percent. High-income countries showed a 10% increase in biodiversity, which was canceled out by a loss in low-income countries. This is despite the fact that high-income countries use five times the ecological resources of low-income countries, which was explained as a result of process whereby wealthy nations are outsourcing resource depletion to poorer nations, which are suffering the greatest ecosystem losses.

Threats

In 2006 many species were formally classified as rare or endangered or threatened; moreover, scientists have estimated that millions more species are at risk which have not been formally recognized. About 40 percent of the 40,177 species assessed using the IUCN Red List criteria are now listed as threatened with extinction—a total of 16,119.

Jared Diamond describes an "Evil Quartet" of habitat destruction, overkill, introduced species and secondary extinctions. Edward O. Wilson prefers the acronym HIPPO, standing for Habitat destruction, Invasive species, Pollution, human over-Population and Over-harvesting. The most authoritative classification in use today is IUCN's Classification of Direct Threats which has been adopted by major international conservation organizations such as the US Nature Conservancy, the World Wildlife Fund, Conservation International and BirdLife International.

Habitat Destruction

Habitat destruction has played a key role in extinctions, especially related to tropical forest destruction. Factors contributing to habitat loss are: overconsumption, overpopulation, land use change, deforestation, pollution (air pollution, water pollution, soil contamination) and global warming or climate change.

Habitat size and numbers of species are systematically related. Physically larger species and those living at lower latitudes or in forests or oceans are more sensitive to reduction in habitat area. Conversion to "trivial" standardized ecosystems (e.g., monoculture following deforestation) effectively destroys habitat for the more diverse species that preceded the conversion. In some countries lack of property rights or lax law/regulatory enforcement necessarily leads to biodiversity loss (degradation costs having to be supported by the community).

A 2007 study conducted by the National Science Foundation found that biodiversity and genetic diversity are codependent—that diversity among species requires diversity within a species and vice versa. "If any one type is removed from the system, the cycle can break down and the community becomes dominated by a single species." At present, the most threatened ecosystems are found in fresh water, according to the Millennium Ecosystem Assessment 2005, which was confirmed by the "Freshwater Animal Diversity Assessment", organised by the biodiversity platform and the French *Institut de recherche pour le développement* (MNHNP).

Co-extinctions are a form of habitat destruction. Co-extinction occurs when the extinction or decline in one accompanies the other, such as in plants and beetles.

Introduced and Invasive Species

Male *Lophura nycthemera* (silver pheasant), a native of East Asia that has been introduced into parts of Europe for ornamental reasons.

Barriers such as large rivers, seas, oceans, mountains and deserts encourage diversity by enabling independent evolution on either side of the barrier, via the process of allopatric speciation. The term invasive species is applied to species that breach the natural barriers that would normally keep them constrained. Without barriers, such species occupy new territory, often supplanting native species by occupying their niches, or by using resources that would normally sustain native species.

The number of species invasions has been on the rise at least since the beginning of the 1900s. Species are increasingly being moved by humans (on purpose and accidentally). In some cases the invaders are causing drastic changes and damage to their new habitats (e.g.: zebra mussels and the emerald ash borer in the Great Lakes region and the lion fish along the North American Atlantic coast). Some evidence suggests that invasive species are competitive in their new habitats because they are subject to less pathogen disturbance. Others report confounding evidence that occasionally suggest that species-rich communities harbor many native and exotic species simultaneously while some say that diverse ecosystems are more resilient and resist invasive plants and animals. An important question is, "do invasive species cause extinctions?" Many studies cite effects of invasive species on natives, but not extinctions. Invasive species seem to increase local (i.e.: alpha diversity) diversity, which decreases turnover of diversity (i.e.: beta diversity). Overall gamma diversity may be lowered because species are going extinct because of other causes, but even some of the most insidious invaders (e.g.: Dutch elm disease, emerald ash borer, chestnut blight in North America) have not caused their host species to become extinct. Extirpation, population decline and homogenization of regional biodiversity are much more common. Human activities have frequently been the cause of invasive species circumventing their barriers, by introducing them for food and other purposes. Human activities therefore allow species to migrate to new areas (and thus become invasive) occurred on time scales much shorter than historically have been required for a species to extend its range.

Not all introduced species are invasive, nor all invasive species deliberately introduced. In cases such as the zebra mussel, invasion of US waterways was unintentional. In other cases, such as mongooses in Hawaii, the introduction is deliberate but ineffective (nocturnal rats were not vulnerable to the diurnal mongoose). In other cases, such as oil palms in Indonesia and Malaysia, the introduction produces substantial economic benefits, but the benefits are accompanied by costly unintended consequences.

Finally, an introduced species may unintentionally injure a species that depends on the species it replaces. In Belgium, Prunus spinosa from Eastern Europe leafs much sooner than its West European counterparts, disrupting the feeding habits of the *Thecla betulae* butterfly (which feeds on the leaves). Introducing new species often leaves endemic and other local species unable to compete with the exotic species and unable to survive. The exotic organisms may be predators, parasites, or may simply outcompete indigenous species for nutrients, water and light.

At present, several countries have already imported so many exotic species, particularly agricultural and ornamental plants, that their own indigenous fauna/flora may be outnumbered. For example, the introduction of kudzu from Southeast Asia to Canada and the United States has threatened biodiversity in certain areas.

Genetic Pollution

Endemic species can be threatened with extinction through the process of genetic pollution, i.e. uncontrolled hybridization, introgression and genetic swamping. Genetic pollution leads to homogenization or replacement of local genomes as a result of either a numerical and/or fitness advantage of an introduced species. Hybridization and introgression are side-effects of introduction and invasion. These phenomena can be especially detrimental to rare species that come into contact with more abundant ones. The abundant species can interbreed with the rare species, swamping its gene pool. This problem is not always apparent from morphological (outward appearance) observations alone. Some degree of gene flow is normal adaptation and not all gene and genotype constellations can be preserved. However, hybridization with or without introgression may, nevertheless, threaten a rare species' existence.

Overexploitation

Overexploitation occurs when a resource is consumed at an unsustainable rate. This occurs on land in the form of overhunting, excessive logging, poor soil conservation in agriculture and the illegal wildlife trade.

About 25% of world fisheries are now overfished to the point where their current biomass is less than the level that maximizes their sustainable yield.

The overkill hypothesis, a pattern of large animal extinctions connected with human migration patterns, can be used explain why megafaunal extinctions can occur within a relatively short time period.

Hybridization, Genetic Pollution/Erosion and Food Security

The Yecoro wheat (right) cultivar is sensitive to salinity, plants resulting from a hybrid cross with cultivar W4910 (left) show greater tolerance to high salinity.

In agriculture and animal husbandry, the Green Revolution popularized the use of conventional hybridization to increase yield. Often hybridized breeds originated in developed countries and were further hybridized with local varieties in the developing world to create high yield strains resistant to local climate and diseases. Local governments and industry have been pushing hybridization. Formerly huge gene pools of various wild and indigenous breeds have collapsed causing widespread genetic erosion and genetic pollution. This has resulted in loss of genetic diversity and biodiversity as a whole.

(GM organisms) have genetic material altered by genetic engineering procedures such as recombinant DNA technology. GM crops have become a common source for genetic pollution, not only of wild varieties but also of domesticated varieties derived from classical hybridization.

Genetic erosion coupled with genetic pollution may be destroying unique genotypes, thereby creating a hidden crisis which could result in a severe threat to our food security. Diverse genetic material could cease to exist which would impact our ability to further hybridize food crops and livestock against more resistant diseases and climatic changes.

Climate Change

Global warming is also considered to be a major potential threat to global biodiversity in the future. For example, coral reefs - which are biodiversity hotspots - will be lost within the century if global warming continues at the current trend.

Polar bears on the sea ice of the Arctic Ocean, near the North Pole. Climate change has started affecting bear populations.

Climate change has seen many claims about potential to affect biodiversity but evidence supporting the statement is tenuous. Increasing atmospheric carbon dioxide certainly affects plant morphology and is acidifying oceans, and temperature affects species ranges, phenology, and weather, but the major impacts that have been predicted are still just *potential* impacts. We have not documented major extinctions yet, even as climate change drastically alters the biology of many species.

In 2004, an international collaborative study on four continents estimated that 10 percent of species would become extinct by 2050 because of global warming. "We need to limit climate change or we wind up with a lot of species in trouble, possibly extinct," said Dr. Lee Hannah, a co-author of the paper and chief climate change biologist at the Center for Applied Biodiversity Science at Conservation International.

A recent study predicts that up to 35% of the world terrestrial carnivores and ungulates will be at higher risk of extinction by 2050 because of the joint effects of predicted climate and land-use change under business-as-usual human development scenarios.

Human Overpopulation

From 1950 to 2011, world population increased from 2.5 billion to 7 billion and is forecast to reach a plateau of more than 9 billion during the 21st century. Sir David King, former chief scientific adviser to the UK government, told a parliamentary inquiry: *"It is self-evident that the massive growth in the human population through the 20th century has had more impact on biodiversity than any other single factor."* At least until the middle of the 21st century, worldwide losses of pristine biodiverse land will probably depend much on the worldwide human birth rate.

According to a 2014 study by the World Wildlife Fund, the global human population already exceeds planet's biocapacity - it would take the equivalent of 1.5 Earths of biocapacity to meet our current demands. The report further points that if everyone on the planet had the Footprint of the average resident of Qatar, we would need 4.8 Earths and if we lived the lifestyle of a typical resident of the USA, we would need 3.9 Earths.

The Holocene Extinction

Rates of decline in biodiversity in this sixth mass extinction match or exceed rates of loss in the five previous mass extinction events in the fossil record. Loss of biodiversity results in the loss of natural capital that supplies ecosystem goods and services. From the perspective of the method known as Natural Economy the economic value of 17 ecosystem services for Earth's biosphere (calculated in 1997) has an estimated value of US$33 trillion (3.3x10[13]) per year.

Conservation

Conservation biology matured in the mid-20th century as ecologists, naturalists and other scientists began to research and address issues pertaining to global biodiversity declines.

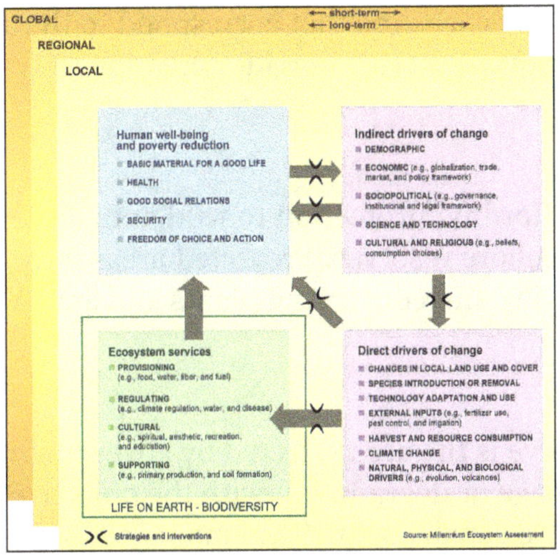

A schematic image illustrating the relationship between biodiversity, ecosystem services, human well-being and poverty. The illustration shows where conservation action, strategies and plans can influence the drivers of the current biodiversity crisis at local, regional, to global scales.

The conservation ethic advocates management of natural resources for the purpose of sustaining biodiversity in species, ecosystems, the evolutionary process and human culture and society.

Conservation biology is reforming around strategic plans to protect biodiversity. Preserving global biodiversity is a priority in strategic conservation plans that are designed to engage public policy and concerns affecting local, regional and global scales of communities, ecosystems and cultures. Action plans identify ways of sustaining human well-being, employing natural capital, market capital and ecosystem services.

In the EU Directive 1999/22/EC zoos are described as having a role in the preservation of the biodiversity of wildlife animals by conducting research or participation in breeding programs.

Protection and Restoration Techniques

Removal of exotic species will allow the species that they have negatively impacted to recover their ecological niches. Exotic species that have become pests can be identified taxonomically (e.g., with Digital Automated Identification SYstem (DAISY), using the barcode of life). Removal is practical only given large groups of individuals due to the economic cost.

As sustainable populations of the remaining native species in an area become assured, "missing" species that are candidates for reintroduction can be identified using databases such as the *Encyclopedia of Life* and the Global Biodiversity Information Facility:

- Biodiversity banking places a monetary value on biodiversity. One example is the Australian Native Vegetation Management Framework.
- Gene banks are collections of specimens and genetic material. Some banks intend to reintroduce banked species to the ecosystem (e.g., via tree nurseries).
- Reduction of and better targeting of pesticides allows more species to survive in agricultural and urbanized areas.
- Location-specific approaches may be less useful for protecting migratory species. One approach is to create wildlife corridors that correspond to the animals' movements. National and other boundaries can complicate corridor creation.

Protected Areas

Protected areas is meant for affording protection to wild animals and their habitat which also includes forest reserves and biosphere reserves. Protected areas have been set up all over the world with the specific aim of protecting and conserving plants and animals.

National Parks

National park and nature reserve is the area selected by governments or private organizations for special protection against damage or degradation with the objective of biodiversity and landscape conservation. National parks are usually owned and managed by national or state governments. A limit is placed on the number of visitors permitted to enter certain fragile areas. Designated trails or roads are created. The visitors are allowed to enter only for study, cultural and recreation purposes. Forestry operations, grazing of animals and hunting of animals are regulated. Exploitation of habitat or wildlife is banned.

Wildlife Sanctuary

Wildlife sanctuary aims only at conservation of species and have the following features:

1. The boundaries of the sanctuaries are not limited by state legislation.
2. The killing, hunting or capturing of any species is prohibited except by or under the control of the highest authority in the department which is responsible for the management of the sanctuary.
3. Private ownership may be allowed.
4. Forestry and other usages can also be permitted.

Forest Reserves

The forests play a vital role in harbouring more than 45,000 floral and 81,000 faunal species of which 5150 floral and 1837 faunal species are endemic. Plant and animal species confined to a specific geographical area are called endemic species. In reserved forests, rights to activities like hunting and grazing are sometimes given to communities living on the fringes of the forest, who sustain their livelihood partially or wholly from forest resources or products. The unclassed forests covers 6.4 percent of the total forest area and they are marked by the following characteristics:

1. They are large inaccessible forests.
2. Many of these are unoccupied.
3. They are ecologically and economically less important.

Steps to Conserve the Forest Cover

1. An extensive reforestation/afforestation program should be followed.
2. Alternative environment-friendly sources of fuel energy such as biogas other than wood should be used.
3. Loss of biodiversity due to forest fire is a major problem, immediate steps to prevent forest fire need to be taken.
4. Overgrazing by cattle can damage a forest seriously. Therefore, certain steps should be taken to prevent overgrazing by cattle.
5. Hunting and poaching should be banned.

Zoological Parks

In zoological parks or zoos, live animals are kept for public recreation, education and conservation purposes. Modern zoos offer veterinary facilities, provide opportunities for threatened species to breed in captivity and usually build environments that simulate the native habitats of the animals in their care. Zoos play a major role in creating awareness among common people about the need to conserve nature.

Botanical Gardens

Botanical garden is a garden in which plants are grown and displayed primarily for scientific and educational purposes. It consists of a collection of living plants, grown outdoors or under glass in greenhouses and conservatories. In addition, it includes a collection of dried plants or herbarium and such facilities as lecture rooms, laboratories, libraries, museums and experimental or research plantings.

Resource Allocation

Focusing on limited areas of higher potential biodiversity promises greater immediate return on investment than spreading resources evenly or focusing on areas of little diversity but greater interest in biodiversity.

A second strategy focuses on areas that retain most of their original diversity, which typically require little or no restoration. These are typically non-urbanized, non-agricultural areas. Tropical areas often fit both criteria, given their natively high diversity and relative lack of development.

Legal Status

A great deal of work is occurring to preserve the natural characteristics of Hopetoun Falls, Australia while continuing to allow visitor access.

International

- United Nations Convention on Biological Diversity (1992) and Cartagena Protocol on Biosafety.
- Convention on International Trade in Endangered Species (CITES).
- Ramsar Convention (Wetlands).
- Bonn Convention on Migratory Species.
- World Heritage Convention (indirectly by protecting biodiversity habitats).
- Regional Conventions such as the Apia Convention.
- Bilateral agreements such as the Japan-Australia Migratory Bird Agreement.

Global agreements such as the Convention on Biological Diversity, give "sovereign national rights over biological resources" (not property). The agreements commit countries to "conserve biodiversity", "develop resources for sustainability" and "share the benefits" resulting from their use. Biodiverse countries that allow bioprospecting or collection of natural products, expect a share of the benefits rather than allowing the individual or institution that discovers/exploits the resource to capture them privately. Bioprospecting can become a type of biopiracy when such principles are not respected.

Sovereignty principles can rely upon what is better known as Access and Benefit Sharing Agreements (ABAs). The Convention on Biodiversity implies informed consent between the source country and the collector, to establish which resource will be used and for what and to settle on a fair agreement on benefit sharing.

National Level Laws

Biodiversity is taken into account in some political and judicial decisions:

- The relationship between law and ecosystems is very ancient and has consequences for biodiversity. It is related to private and public property rights. It can define protection for threatened ecosystems, but also some rights and duties (for example, fishing and hunting rights).
- Law regarding species is more recent. It defines species that must be protected because they may be threatened by extinction. The U.S. Endangered Species Act is an example of an attempt to address the "law and species" issue.
- Laws regarding gene pools are only about a century old. Domestication and plant breeding methods are not new, but advances in genetic engineering have led to tighter laws covering distribution of genetically modified organisms, gene patents and process patents. Governments struggle to decide whether to focus on for example, genes, genomes, or organisms and species.

Uniform approval for use of biodiversity as a legal standard has not been achieved, however. Bosselman argues that biodiversity should not be used as a legal standard, claiming that the remaining areas of scientific uncertainty cause unacceptable administrative waste and increase litigation without promoting preservation goals.

India passed the Biological Diversity Act in 2002 for the conservation of biological diversity in India. The Act also provides mechanisms for equitable sharing of benefits from the use of traditional biological resources and knowledge.

Analytical Limits

Taxonomic and Size Relationships

Less than 1% of all species that have been described have been studied beyond simply noting their existence. The vast majority of Earth's species are microbial. Contemporary biodiversity physics is "firmly fixated on the visible [macroscopic] world". For example, microbial life is metabolically and environmentally more diverse than multicellular life (see e.g., extremophile). "On the tree of life, based on analyses of small-subunit ribosomal RNA, visible life consists of barely noticeable twigs. The inverse relationship of size and population recurs higher on the evolutionary ladder—"to a first approximation, all multicellular species on Earth are insects". Insect extinction rates are high—supporting the Holocene extinction hypothesis.

Ecosystem Services

Humankind benefits in a multitude of ways from ecosystems. Collectively, these benefits are becoming known as ecosystem services. Ecosystem services are regularly involved in the provisioning of clean drinking water and the decomposition of wastes. While scientists and environmentalists have discussed ecosystem services implicitly for decades, the ecosystem services concept itself was popularized by the Millennium Ecosystem Assessment (MA) in the early 2000s. This grouped ecosystem services into four broad categories: *provisioning*, such as the production of food and water; *regulating*, such as the control of climate and disease; *supporting*, such as nutrient cycles and crop pollination; and *cultural,* such as spiritual and recreational benefits. To help inform decision-makers, many ecosystem services are being assigned economic values.

Honey bee on Avocado crop. Pollination is just one type of ecosystem service.

History

While the notion of human dependence on Earth's ecosystems reaches to the start of homo sapiens' existence, the term 'natural capital' was first coined by E.F. Schumacher in 1973 in his book *Small is Beautiful* . Recognition of how ecosystems could provide more complex services to mankind date back to at least Plato (c. 400 BC) who understood that deforestation could lead to soil erosion and the drying of springs. Modern ideas of ecosystem services probably began with Marsh in 1864 when he challenged the idea that Earth's natural resources are unbounded by pointing out changes in soil fertility in the Mediterranean. It was not until the late 1940s that three key authors – Henry Fairfield Osborn, Jr, William Vogt, and Aldo Leopold – promoted recognition of human dependence on the environment.

In 1956, Paul Sears drew attention to the critical role of the ecosystem in processing wastes and recycling nutrients. In 1970, Paul Ehrlich and Rosa Weigert called attention to "ecological systems" in their environmental science textbook and "the most subtle and dangerous threat to man's existence... the potential destruction, by man's own activities, of those ecological systems upon which the very existence of the human species depends".

The term "environmental services" was introduced in a 1970 report of the *Study of Critical Environmental Problems,* which listed services including insect pollination, fisheries, climate regulation and flood control. In following years, variations of the term were used, but eventually 'ecosystem services' became the standard in scientific literature.

The ecosystem services concept has continued to expand and includes socio-economic and conservation objectives, which are discussed below. A history of the concepts and terminology of ecosystem services as of 1997, can be found in Daily's book "Nature's Services: Societal Dependence on Natural Ecosystems".

Definition

Per the 2006 Millennium Ecosystem Assessment (MA), ecosystem services are "the benefits people obtain from ecosystems." The MA also delineated the four categories of ecosystem services—

supporting, provisioning, regulating and cultural—discussed below.

By 2010, there had evolved various working definitions and descriptions of ecosystem services in the literature. To prevent double counting in ecosystem services audits, for instance, The Economics of Ecosystems and Biodiversity (TEEB) replaced "Supporting Services" in the MA with "Habitat Services" and "ecosystem functions," defined as "a subset of the interactions between ecosystem structure and processes that underpin the capacity of an ecosystem to provide goods and services."

Four Categories

The Millennium Ecosystem Assessment (MA) report 2005 defines *Ecosystem services* as benefits people obtain from ecosystems and distinguishes four categories of ecosystem services, where the so-called supporting services are regarded as the basis for the services of the other three categories. The following lists represent the definition and samples of each according to the MA:

Supporting Services

Ecosystem services "that are necessary for the production of all other ecosystem services". These include services such as nutrient recycling, primary production and soil formation. These services make it possible for the ecosystems to provide services such as food supply, flood regulation and water purification.

Provisioning Services

"Products obtained from ecosystems":

- Food (including seafood and game), crops, wild foods, and spices.
- Raw materials (including lumber, skins, fuel wood, organic matter, fodder, and fertilizer).
- Genetic resources (including crop improvement genes, and health care).
- Water.
- Minerals (including diatomite).
- Medicinal resources (including pharmaceuticals, chemical models, and test and assay organisms).
- Energy (hydropower, biomass fuels).
- Ornamental resources (including fashion, handicraft, jewelry, pets, worship, decoration and souvenirs like furs, feathers, ivory, orchids, butterflies, aquarium fish, shells, etc.).

Regulating Services

"Benefits obtained from the regulation of ecosystem processes":

- Carbon sequestration and climate regulation.

- Waste decomposition and detoxification.
- Purification of water and air.
- Pest and disease control.

Cultural Services

"Nonmaterial benefits people obtain from ecosystems through spiritual enrichment, cognitive development, reflection, recreation, and aesthetic experiences":

- cultural (including use of nature as motif in books, film, painting, folklore, national symbols, architect, advertising, etc.).
- spiritual and historical (including use of nature for religious or heritage value or natural).
- recreational experiences (including ecotourism, outdoor sports, and recreation).
- science and education (including use of natural systems for school excursions, and scientific discovery).
- Therapeutic (including Ecotherapy, social forestry and animal assisted therapy).

There is discussion as to how the concept of cultural ecosystem services can be operationalized. A good review of approaches in landscape aesthetics, cultural heritage, outdoor recreation, and spiritual significance to define and assess cultural values of our environment so that they fit into the ecosystem services approach is given by Daniel et al. who vote for models that explicitly link ecological structures and functions with cultural values and benefits. There also is a fundamental critique of the concept of cultural ecosystem services that builds on three arguments:

1. Pivotal cultural values attaching to the natural/cultivated environment rely on an area's unique character that cannot be addressed by methods that use universal scientific parameters to determine ecological structures and functions.
2. If a natural/cultivated environment has symbolic meanings and cultural values the object of these values are not ecosystems but shaped lifeworldly phenomena like mountains, lakes, forests, and, mainly, symbolic landscapes.
3. Those cultural values do result not from properties produced by ecosystems but are the product of a specific way of seeing within the given cultural framework of symbolic experience.

Examples

The following examples illustrate the relationships between humans and natural ecosystems through the services derived from them:

- In New York City, where the quality of drinking water had fallen below standards required by the U.S. Environmental Protection Agency (EPA), authorities opted to restore the polluted Catskill Watershed that had previously provided the city with the ecosystem service of water purification. Once the input of sewage and pesticides to the watershed area was reduced, natural abiotic processes such as soil absorption and filtration of chemicals, together with biotic recycling via root systems and soil microorganisms, water quality improved to levels that met government standards. The cost of this investment in natural capital was estimated between $1–1.5 billion, which contrasted dramatically with the estimated $6–8 billion cost of constructing a water filtration plant plus the $300 million annual running costs.

- Pollination of crops by bees is required for 15-30% of U.S. food production; most large-scale farmers import non-native honey bees to provide this service. One study reports that in California's agricultural region, it was found that wild bees alone could provide partial or complete pollination services or enhance the services provided by honey bees through behavioral interactions. However, intensified agricultural practices can quickly erode pollination services through the loss of species and those remaining are unable to compensate for the difference. The results of this study also indicate that the proportion of chaparral and oak-woodland habitat available for wild bees within 1–2 km of a farm can strongly stabilize and enhance the provision of pollination services, thereby providing a potential insurance policy for farmers of this region.
- In watersheds of the Yangtze River (China), spatial models for water flow through different forest habitats were created to determine potential contributions for hydroelectric power in the region. By quantifying the relative value of ecological parameters (vegetation-soil-slope complexes), researchers were able to estimate the annual economic benefit of maintaining forests in the watershed for power services to be 2.2 times that if it were harvested once for timber.
- In the 1980s, mineral water company Vittel (now a brand of Nestlé Waters) faced a critical problem. Nitrates and pesticides were entering the company's springs in northeastern France. Local farmers had intensified agricultural practices and cleared native vegetation that previously had filtered water before it seeped into the aquifer used by Vittel. This contamination threatened the company's right to use the "natural mineral water" label under French law. In response to this business risk, Vittel developed an incentive package for farmers to improve their agricultural practices and consequently reduce water pollution that had affected Vittel's product. For example, Vittel provided subsidies and free technical assistance to farmers in exchange for farmers' agreement to enhance pasture management, reforest catchments, and reduce the use of agrochemicals. This is an example of a Payment for ecosystem services program.
- It was counted that to plant 15 000 ha new woodland in UK, if we consider only the value of timber, it would cost £ 79 000 000 which is more than the benefit of £ 65 000 000. If, however, we include all other benefits the trees in lowland could provide (like soil stabilization, wind deflection, recreation wind defense, food production, air purification, carbon storage, wildlife habitat, fuel production, cooling, flood prevention), the costs will increase due to displacing the profitable farmland (would be around £ 231 000 000) but will be overweight by benefits of £ 546 000 000.
- In Europe, various projects are implemented in order to define the values of concrete eco-

systems and to implement this concept into decision making process. For example, "LIFE Viva grass" project aims to due this with grasslands in Baltics.

Ecology

Understanding of ecosystem services requires a strong foundation in ecology, which describes the underlying principles and interactions of organisms and the environment. Since the scales at which these entities interact can vary from microbes to landscapes, milliseconds to millions of years, one of the greatest remaining challenges is the descriptive characterization of energy and material flow between them. For example, the area of a forest floor, the detritus upon it, the microorganisms in the soil and characteristics of the soil itself will all contribute to the abilities of that forest for providing ecosystem services like carbon sequestration, water purification, and erosion prevention to other areas within the watershed. Note that it is often possible for multiple services to be bundled together and when benefits of targeted objectives are secured, there may also be ancillary benefits – the same forest may provide habitat for other organisms as well as human recreation, which are also ecosystem services.

The complexity of Earth's ecosystems poses a challenge for scientists as they try to understand how relationships are interwoven among organisms, processes and their surroundings. As it relates to human ecology, a suggested research agenda for the study of ecosystem services includes the following steps:

1. Identification of *ecosystem service providers* (*ESPs*) – species or populations that provide specific ecosystem services – and characterization of their functional roles and relationships.
2. Determination of community structure aspects that influence how ESPs function in their natural landscape, such as compensatory responses that stabilize function and non-random extinction sequences which can erode it.
3. Assessment of key environmental (abiotic) factors influencing the provision of services.
4. Measurement of the spatial and temporal scales ESPs and their services operate on.

Recently, a technique has been developed to improve and standardize the evaluation of ESP functionality by quantifying the relative importance of different species in terms of their efficiency and abundance. Such parameters provide indications of how species respond to changes in the environment (i.e. predators, resource availability, climate) and are useful for identifying species that are disproportionately important at providing ecosystem services. However, a critical drawback is that the technique does not account for the effects of interactions, which are often both complex and fundamental in maintaining an ecosystem and can involve species that are not readily detected as a priority. Even so, estimating the functional structure of an ecosystem and combining it with information about individual species traits can help us understand the resilience of an ecosystem amidst environmental change.

Many ecologists also believe that the provision of ecosystem services can be stabilized with biodiversity. Increasing biodiversity also benefits the variety of ecosystem services available to society. Understanding the relationship between biodiversity and an ecosystem's stability is essential to the management of natural resources and their services.

Redundancy Hypothesis

The concept of ecological redundancy is sometimes referred to as *functional compensation* and assumes that more than one species performs a given role within an ecosystem. More specifically, it is characterized by a particular species increasing its efficiency at providing a service when conditions are stressed in order to maintain aggregate stability in the ecosystem. However, such increased dependence on a compensating species places additional stress on the ecosystem and often enhances its susceptibility to subsequent disturbance. The redundancy hypothesis can be summarized as "species redundancy enhances ecosystem resilience".

Another idea uses the analogy of rivets in an airplane wing to compare the exponential effect the loss of each species will have on the function of an ecosystem; this is sometimes referred to as *rivet popping*. If only one species disappears, the loss of the ecosystem's efficiency as a whole is relatively small; however if several species are lost, the system essentially collapses as an airplane wing would, were it to lose too many rivets. The hypothesis assumes that species are relatively specialized in their roles and that their ability to compensate for one another is less than in the redundancy hypothesis. As a result, the loss of any species is critical to the performance of the ecosystem. The key difference is the rate at which the loss of species affects total ecosystem function.

Portfolio Effect

A third explanation, known as the *portfolio effect*, compares biodiversity to stock holdings, where diversification minimizes the volatility of the investment, or in this case, the risk in stability of ecosystem services. This is related to the idea of *response diversity* where a suite of species will exhibit differential responses to a given environmental perturbation and therefore when considered together, they create a stabilizing function that preserves the integrity of a service.

Several experiments have tested these hypotheses in both the field and the lab. In ECOTRON, a laboratory in the UK where many of the biotic and abiotic factors of nature can be simulated, studies have focused on the effects of earthworms and symbiotic bacteria on plant roots. These laboratory experiments seem to favor the rivet hypothesis. However, a study on grasslands at Cedar Creek Reserve in Minnesota seems to support the redundancy hypothesis, as have many other field studies.

Economics

There are questions regarding the environmental and economic values of ecosystem services. Some people may be unaware of the environment in general and humanity's interrelatedness with the natural environment, which may cause misconceptions. Although environmental awareness is rapidly improving in our contemporary world, ecosystem capital and its flow are still poorly understood, threats continue to impose, and we suffer from the so-called 'tragedy of the commons'. Many efforts to inform decision-makers of current versus future costs and benefits now involve organizing and translating scientific knowledge to economics, which articulate the consequences of our choices in comparable units of impact on human well-being. An especially challenging aspect of this process is that interpreting ecological information collected from one spatial-temporal scale does not necessarily mean it can be applied at another; understanding the dynamics of ecological processes relative to ecosystem services is essential in aiding economic decisions. Weighting factors such as a service's irreplaceability or bundled services can also allocate economic value such that goal attainment becomes more efficient.

Sustainable urban drainage pond near housing in Scotland. The filtering and cleaning of surface and waste water by natural vegetation is a form of ecosystem service.

The economic valuation of ecosystem services also involves social communication and information, areas that remain particularly challenging and are the focus of many researchers. In general, the idea is that although individuals make decisions for any variety of reasons, trends reveal the aggregative preferences of a society, from which the economic value of services can be inferred and assigned. The six major methods for valuing ecosystem services in monetary terms are:

- Avoided cost: Services allow society to avoid costs that would have been incurred in the absence of those services (e.g. waste treatment by wetland habitats avoids health costs).
- Replacement cost: Services could be replaced with man-made systems (e.g. restoration of the Catskill Watershed cost less than the construction of a water purification plant).
- Factor income: Services provide for the enhancement of incomes (e.g. improved water quality increases the commercial take of a fishery and improves the income of fishers).
- Travel cost: Service demand may require travel, whose costs can reflect the implied value of the service (e.g. value of ecotourism experience is at least what a visitor is willing to pay to get there).
- Hedonic pricing: Service demand may be reflected in the prices people will pay for associated goods (e.g. coastal housing prices exceed that of inland homes).
- Contingent valuation: Service demand may be elicited by posing hypothetical scenarios that involve some valuation of alternatives (e.g. visitors willing to pay for increased access to national parks).

A peer-reviewed study published in 1997 estimated the value of the world's ecosystem services and natural capital to be between US$16–54 trillion per year, with an average of US$33 trillion per year. However, Salles (2011) indicates 'The total value of biodiversity is infinite, so having debate about what is the total value of nature is actually pointless because we can't live without it'.

Management and Policy

Although monetary pricing continues with respect to the valuation of ecosystem services, the challenges in policy implementation and management are significant and multitudinous. The administration of common pool resources is a subject of extensive academic pursuit. From defining the problems to finding solutions that can be applied in practical and sustainable ways, there is much to overcome. Considering options must balance present and future human needs, and decision-makers must frequently work from valid but incomplete information. Existing legal policies are often considered insufficient since they typically pertain to human health-based standards that are mismatched with necessary means to protect ecosystem health and services. To improve the information available, one suggestion has involved the implementation of an *Ecosystem Services Framework* (ESF), which integrates the biophysical and socio-economic dimensions of protecting the environment and is designed to guide institutions through multidisciplinary information and jargon, helping to direct strategic choices.

Novel and expedient methods are needed to deal with managing Earth's ecosystem services. Local to regional collective management efforts might be considered appropriate for services like crop pollination or resources like water. Another approach that has become increasingly popular over the last decade is the marketing of ecosystem services protection. Payment and trading of services is an emerging worldwide small-scale solution where one can acquire credits for activities such as sponsoring the protection of carbon sequestration sources or the restoration of ecosystem service providers. In some cases, banks for handling such credits have been established and conservation companies have even gone public on stock exchanges, defining an evermore parallel link with economic endeavors and opportunities for tying into social perceptions. However, crucial for implementation are clearly defined land rights, which is often lacking in many developing countries. In particular, many forest-rich developing countries suffering deforestation experience conflict between different forest stakeholders. In addition, concerns for such global transactions include inconsistent compensation for services or resources sacrificed elsewhere and misconceived warrants for irresponsible use. Another approach has been focused on protecting ecosystem service 'hotspots'. Recognition that the conservation of many ecosystem services aligns with more traditional conservation goals (i.e. biodiversity) has led to the suggested merging of objectives for maximizing their mutual success. This may be particularly strategic when employing networks that permit the flow of services across landscapes, and might also facilitate securing the financial means to protect services through a diversification of investors.

For example, in recent years there has been interest in the valuation of ecosystem services provided by shellfish production and restoration. A keystone species, low in the food chain, bivalve shellfish such as oysters support a complex community of species by performing a number of functions essential to the diverse array of species that surround them. There is also increasing recognition that some shellfish species may impact or control many ecological processes; so much so that they are included on the list of "ecosystem engineers"—organisms that physically, biologically or chemically modify the environment around them in ways that influence the health of other organisms. Many of the ecological functions and processes performed or affected by shellfish contribute to human well-being by providing a stream of valuable ecosystem services over time by filtering out particulate materials and potentially mitigating water quality issues by controlling excess nutrients in the water.

Ecosystem-Based Adaptation (EbA)

Ecosystem-Based Adaptation or EbA is an emerging strategy for community development and environmental management that seeks to use an ecosystem services framework to help communities adapt to the effects of climate change. The Convention on Biological Diversity currently defines Ecosystem-Based Adaptation as "the use of biodiversity and ecosystem services to help people adapt to the adverse effects of climate change", which includes the use of "sustainable management, conservation and restoration of ecosystems, as part of an overall adaptation strategy that takes into account the multiple social, economic and cultural co-benefits for local communities".

In 2001, the Millennium Ecosystem Assessment announced that humanity's impact on the natural world was increasing to levels never before seen, and that the degradation of the planet's ecosystems would become a major barrier to achieving the Millennium Development Goals. In recognition of this fact, Ecosystem-Based Adaptation seeks to use the restoration of ecosystems as a stepping-stone to improving the quality of life in communities experiencing the impacts of climate change. Specifically, this involves the restoration of ecosystems that provide the community with essential services, such as the provisioning of food and water and protection from storm surges and flooding. EbA interventions typically combine elements of both climate change mitigation and adaptation to global warming to help address the community's current and future needs.

Collaborative planning between scientists, policy makers, and community members is an essential element of Ecosystem-Based Adaptation. By drawing on the expertise of outside experts and local residents alike, EbA seeks to develop unique solutions to unique problems, rather than simply replicating past projects.

Estuarine and Coastal Ecosystem Services

Ecosystem services are defined as the gains acquired by humankind from surroundings ecosystems. Four different types of ecosystem services have been distinguished by the scientific body: regulating services, provisioning services, cultural services and supporting services. An ecosystem does not necessarily offer all four types of services simultaneously; but given the intricate nature of any ecosystem, it is usually assumed that humans benefit from a combination of these services. The services offered by diverse types of ecosystems (forests, seas, coral reefs, mangroves, etc.) differ in nature and in consequence. In fact, some services directly affect the livelihood of neighboring human populations (such as fresh water, food or aesthetic value, etc.) while other services affect general environmental conditions by which humans are indirectly impacted (such as climate change, erosion regulation or natural hazard regulation, etc.).

Estuarine and coastal ecosystems are both marine ecosystems. An estuary is defined as the area in which a river meets the sea or the ocean. The waters surrounding this area are predominantly salty waters or brackish waters; and the incoming river water is dynamically motioned by the tide. An estuary strip may be covered by populations of reed (or similar plants) and/or sandbanks (or similar form or land).

A coastal ecosystem occurs in areas where the sea or ocean waters meet the land.

Regulating Services

Regulating services are the "benefits obtained from the regulation of ecosystem processes". In the case of coastal and estuarine ecosystems, these services include climate regulation, waste treatment and disease control and natural hazard regulation.

- Climate Regulation:

Both the biotic and abiotic ensembles of marine ecosystems play a role in climate regulation. They act as sponges when it comes to gases in the atmosphere, retaining large levels of CO2 and other Green House Gases (methane and nitrous oxide). Marine plants also use CO2 for photosynthesis purposes and help in reducing the atmospheric CO2. The oceans and seas absorb the heat from the atmosphere and redistribute it through the means of water currents, and atmospheric processes, such as evaporation and the reflection of light allow for the cooling and warming of the overlying atmosphere. The ocean temperatures are thus imperative to the regulation of the atmospheric temperatures in any part of the world: "without the ocean, the Earth would be unbearably hot during the daylight hours and frigidly cold, if not frozen, at night".

- Waste Treatment & Disease Regulation:

Another service offered by marine ecosystem is the treatment of wastes, thus helping in the regulation of diseases. Wastes can be diluted and detoxified through transport across marine ecosystems; pollutants are removed from the environment and stored, buried or recycled in marine ecosystems: "Marine ecosystems break down organic waste through microbial communities that filter water, reduce/limit the effects of eutrophication, and break down toxic hydrocarbons into their basic components such as carbon dioxide, nitrogen, phosphorus, and water". The fact that waste is diluted with large volumes of water and moves with water currents leads to the regulation of diseases and the reduction of toxics in seafood.

- Buffer Zones:

Coastal and estuarine ecosystems act as buffer zones against natural hazards and environmental disturbances, such as floods, cyclones, tidal surges and storms. The role they play is to "[absorb] a portion of the impact and thus [lessen] its effect on the land". Wetlands, for example, and the vegetation it supports – trees, root mats, etc. – retain large amounts of water (surface water, snowmelt, rain, groundwater) and then slowly releases them back, decreasing the likeliness of floods. Mangrove forests protect coastal shorelines from tidal erosion or erosion by currents; a process that was studied after the 1999 cyclone that hit India. Villages that were surrounded with mangrove forests encountered less damages than other villages that weren't protected by mangroves.

Provisioning Services

Provisioning services consist of all "the products obtained from ecosystems". Marine ecosystems provide people with: wild & cultured seafood, fresh water, fiber & fuel and biochemical & genetic resources.

- Marine Products

Humans consume a large number of products originating from the seas, whether as a nutritious product or for use in other sectors: "More than one billion people worldwide, or one-sixth of the global population, rely on fish as their main source of animal protein. In 2000, marine and coastal fisheries accounted for 12 per cent of world food production". Fish and other edible marine products - primarily fish, shellfish, roe and seaweeds – constitute for populations living along the coast the main elements of the local cultural diets, norms and traditions. A very pertinent example would be Sushi, the national food of Japan, which consists mostly of different types of fish and seaweed.

- Fresh Water

Water bodies that are not highly concentrated in salts are referred to as 'fresh water' bodies. Fresh water may run through lakes, rivers and streams, to name a few; but it is most prominently found in the frozen state or as soil moisture or buried deep underground. Fresh water is not only important for the survival of humans, but also for the survival of all the existing species of animals, plants.

- Raw Materials

Marine creatures provide us with the raw materials needed for the manufacturing of clothing, building materials (lime extracted from coral reefs), ornamental items and personal-use items (luffas, art and jewelry): "The skin of marine mammals for clothing, gas deposits for energy production, lime (extracted from coral reefs) for building construction, and the timber of mangroves and coastal forests for shelter are some of the more familiar uses of marine organisms. Raw marine materials are utilized for non-essential goods as well, such as shells and corals in ornamental items". Humans have also referred to processes within marine environments for the production of renewable energy: using the power of waves – or tidal power – as a source of energy for the powering of a turbine, for example. Oceans and seas are used as sites for offshore oil and gas installations, offshore wind farms.

- Biochemical and Genetic Resources

Biochemical resources are compounds extracted from marine organisms for use in medicines, pharmaceuticals, cosmetics and other biochemical products. Genetic resources are the genetic information found in marine organisms that would later on be used for animal and plant breeding and for technological advances in the biological field. These resources are either directly taken out from an organism – such as fish oil as a source of omega3 –, or used as a model for innovative man-made products: "such as the construction of fiber optics technology based on the properties of sponges. [...] Compared to terrestrial products, marine-sourced products tend to be more highly bioactive, likely due to the fact that marine organisms have to retain their potency despite being diluted in the surrounding sea-water".

Cultural Services

Cultural services relate to the non-material world, as they benefit the benefit recreational, aesthetic, cognitive and spiritual activities, which are not easily quantifiable in monetary terms.

- Inspirational

Marine environments have been used by many as an inspiration for their works of art, music, architecture, traditions... Water environments are spiritually important as a lot of people view them as a means for rejuvenation and change of perspective. Many also consider the water as being a part of their personality, especially if they have lived near it since they were kids: they associate it to fond memories and past experiences. Living near water bodies for a long time results in a certain set of water activities that become a ritual in the lives of people and of the culture in the region.

- Recreation and Tourism

Sea sports are very popular among coastal populations: surfing, snorkeling, whale watching, kayaking, recreational fishing...a lot of tourists also travel to resorts close to the sea or rivers or lakes to be able to experience these activities, and relax near the water.

Beach accommodated into a recreational area.

- Science and Education

A lot can be learned from marine processes, environments and organisms – that could be implemented into our daily actions and into the scientific domain. Although much is still yet to still be known about the ocean world: "by the extraordinary intricacy and complexity of the marine environment and how it is influenced by large spatial scales, time lags, and cumulative effects".

Supporting Services

Supporting services are the services that allow for the other ecosystem services to be present. They have indirect impacts on humans that last over a long period of time. Several services can be considered as being both supporting services and regulating/cultural/provisioning services.

- Nutrient Cycling

"Nutrient cycling refers to the storage, cycling, and maintenance of nutrients by organisms and their associated processes". The ocean is a vast storage pool for these nutrients, such as carbon, nitrogen and phosphorus. The nutrients are absorbed by the basic organisms of the marine food web and are thus transferred from one organism to the other and from one ecosystem to the other. Nutrients are recycled through the life cycle of organisms as they die and decompose, releasing the nutrients into the neighboring environment. "The service of nutrient cycling eventually impacts all other ecosystem services as all living things require a constant supply of nutrients to survive".

- Biologically Mediated Habitats

Biologically mediated habitats are defined as being the habitats that living marine structures offer to other organisms. These need not to be designed for the sole purpose of serving as a habitat, but happen to become living quarters whilst growing naturally. For example, coral reefs and mangrove forests are home to numerous species of fish, seaweed and shellfish... The importance of these habitats is that they allow for interactions between different species, aiding the provisioning of marine goods and services. They are also very important for the growth at the early life stages of marine species (breeding and bursary spaces), as they serve as a food source and as a shelter from predators.

Coral and other living organisms serve as habitats for many marine species.

- Primary Production

Primary production refers to the production of organic matter, i.e., chemically bound energy, through processes such as photosynthesis and chemosynthesis. The organic matter produced by primary producers forms the basis of all food webs. Further, it generates oxygen, a molecule necessary for life.

Businessworld

Ecosystem services degradation can pose a number of risks to corporate performance as well as provide business opportunities through ecosystem restoration and enhancement. Risks and opportunities include:

- Operational:
    - Risks such as higher costs for freshwater due to scarcity or lower output for hydroelectric facilities due to siltation.
    - Opportunities such as increasing water-use efficiency or building an on-site wetland to circumvent the need for new water treatment infrastructure.
- Regulatory and legal:
    - Risks such as new fines, government regulations, or lawsuits from local communities that lose ecosystem services due to corporate activities.
    - Opportunities such as engaging governments to develop policies and incentives to protect or restore ecosystems that provide services a company needs.
- Reputational:

    - Risks such as retail companies being targeted by nongovernmental organization campaigns for purchasing wood or paper from sensitive forests.

  Opportunities such as implementing and communicating sustainable purchasing, operating, or investment practices in order to differentiate corporate brands.

- Market and product:
    - Risks such as customers switching to other suppliers that offer products with lower ecosystem impacts or governments implementing new sustainable procurement policies.
    - Opportunities such as launching new products and services that reduce customer impacts on ecosystems or participating in emerging markets for carbon sequestration and watershed protection other products.
- Financing:
    - Risks such as banks implementing more rigorous lending requirements for corporate loans.
    - Opportunities such as banks offering more favorable loan terms or investors taking positions in companies supplying products and services that improve resource use efficiency or restore degraded ecosystems.

Many companies are not fully aware of the extent of their dependence and impact on ecosystems and the possible ramifications. Likewise, environmental management systems and environmental due diligence tools are more suited to handle "traditional" issues of pollution and natural resource consumption. Most focus on environmental impacts, not dependence. Several newly developed tools and methodologies can help the private sector value and assess ecosystem services. These include Our Ecosystem, the Corporate Ecosystem Services Review (ESR), Artificial Intelligence for Ecosystem Services (ARIES), the Natural Value Initiative (NVI) and InVEST (Integrated Valuation of Ecosystem Services & Tradeoffs)

Land use Change Decisions

Ecosystem services decisions require making complex choices at the intersection of ecology, technology, society and the economy. The process of making ecosystem services decisions must consider the interaction of many types of information, honor all stakeholder viewpoints, including regulatory agencies, proposal proponents, decision makers, residents, NGOs, and measure the impacts on all four parts of the intersection. These decisions are usually spatial, always multi-objective, and based on uncertain data, models, and estimates. Often it is the combination of the best science combined with the stakeholder values, estimates and opinions that drive the process.

One analytical study modeled the stakeholders as agents to support water resource management decisions in the Middle Rio Grande basin of New Mexico. This study focused on modeling the stakeholder inputs across a spatial decision, but ignored uncertainty. Another study used Monte Carlo methods to exercise econometric models of landowner decisions in a study of the effects of land-use change. Here the stakeholder inputs were modeled as random effects to reflect the uncertainty. A third study used a Bayesian decision support system to both model the uncertainty in the scientific information Bayes Nets and to assist collecting and fusing the input from stakeholders. This study was about siting wave energy devices off the Oregon Coast, but presents a general method for managing uncertain spatial science and stakeholder information in a decision making environment. Remote sensing data and analyses can be used to assess the health and extent of land cover classes that provide ecosystem services, which aids in planning, management, monitoring of stakeholders' actions, and communication between stakeholders.

In Baltic countries scientists, nature conservationists and local authorities are implementing integrated planning approach for grassland ecosystems. They are develping Integrated Planning Tool that will be based on GIS (geographic information system) technology and put online that will help for planners to choose the best grassland management solution for concrete grassland. It will look holistically at the processes in the countryside and help to find best grassland management solutions by taking into account both natural and socioeconomic factors of the particular site.

Environmentalism

Environmentalism or environmental rights is a broad philosophy, ideology, and social movement regarding concerns for environmental protection and improvement of the health of the environment, particularly as the measure for this health seeks to incorporate the concerns of non-human elements. Environmentalism advocates the lawful preservation, restoration and/or improvement of the natural environment, and may be referred to as a movement to control pollution or protect plant and animal diversity. For this reason, concepts such as a land ethic, environmental ethics, biodiversity, ecology, and the biophilia hypothesis figure predominantly.

At its crux, environmentalism is an attempt to balance relations between humans and the various natural systems on which they depend in such a way that all the components are accorded a proper degree of sustainability. The exact measures and outcomes of this balance is controversial and there are many different ways for environmental concerns to be expressed in practice. Environmentalism and environmental concerns are often represented by the color green, but this association has been appropriated by the marketing industries for the tactic known as greenwashing. Environmentalism is opposed by anti-environmentalism, which says that the Earth is less fragile than some environmentalists maintain, and portrays environmentalism as overreacting to the human contribution to climate change or opposing human advancement.

Definitions

Environmentalism denotes a social movement that seeks to influence the political process by lobbying, activism, and education in order to protect natural resources and ecosystems. The word was first coined in 1922.

An *environmentalist* is a person who may speak out about our natural environment and the sustainable management of its resources through changes in public policy or individual behavior. This may include supporting practices such as informed consumption, conservation initiatives, investment in renewable resources, improved efficiencies in the materials economy, transitioning to new accounting paradigms such as Ecological economics and renewing and revitalizing our connections with non-human life.

In various ways (for example, grassroots activism and protests), environmentalists and environmental organizations seek to give the natural world a stronger voice in human affairs.

In general terms, environmentalists advocate the sustainable management of resources, and the protection (and restoration, when necessary) of the natural environment through changes in public policy and individual behavior. In its recognition of humanity as a participant in ecosystems, the movement is centered around ecology, health, and human rights.

While the term *environmentalism* focuses more on the environmental and nature-related aspects of green ideology and politics, *ecologism* as a term combines the ideology of social ecology and environmentalism. *Ecologism* as a term is more commonly used in continental European languages while *environmentalism* is more commonly used in English but the words have slightly different connotations.

History

A concern for environmental protection has recurred in diverse forms, in different parts of the world, throughout history. For example, in Europe, King Edward I of England banned the burning of sea-coal by proclamation in London in 1272, after its smoke had become a problem. The fuel was so common in England that this earliest of names for it was acquired because it could be carted away from some shores by the wheelbarrow.

Earlier in the Middle East, the Caliph Abu Bakr in the 630s commanded his army to "Bring no harm to the trees, nor burn them with fire," and "Slay not any of the enemy's flock, save for your food." Arabic medical treatises during the 9th to 13th centuries dealing with environmentalism and environmental science, including pollution, were written by Al-Kindi, Qusta ibn Luqa, Al-Razi, Ibn Al-Jazzar, al-Tamimi, al-Masihi, Avicenna, Ali ibn Ridwan, Ibn Jumay, Isaac Israeli ben Solomon, Abd-el-latif, Ibn al-Quff, and Ibn al-Nafis. Their works covered a number of subjects related to pollution, such as air pollution, water pollution, soil contamination, municipal solid waste mishandling, and environmental impact assessments of certain localities.

Early Environmental Legislation

The origins of the environmental movement lay in the response to increasing levels of smoke pollution in the atmosphere during the Industrial Revolution. The emergence of great factories and the concomitant immense growth in coal consumption gave rise to an unprecedented level of air pollution in industrial centers; after 1900 the large volume of industrial chemical discharges added to the growing load of untreated human waste. The first large-scale, modern environmental laws came in the form of Britain's Alkali Acts, passed in 1863, to regulate the deleterious air pollution (gaseous hydrochloric acid) given off by the Leblanc process, used to produce soda ash. An Alkali inspector and four sub-inspectors were appointed to curb this pollution. The responsibilities of the inspectorate were gradually expanded, culminating in the Alkali Order 1958 which placed all major heavy industries that emitted smoke, grit, dust and fumes under supervision.

Levels of air pollution rose during the Industrial Revolution, sparking the first modern environmental laws to be passed in the mid-19th century.

In industrial cities local experts and reformers, especially after 1890, took the lead in identifying environmental degradation and pollution, and initiating grass-roots movements to demand and achieve reforms. Typically the highest priority went to water and air pollution. The Coal Smoke Abatement Society was formed in 1898 making it one of the oldest environmental NGOs. It was founded by artist Sir William Blake Richmond, frustrated with the pall cast by coal smoke. Although there were earlier pieces of legislation, the Public Health Act 1875 required all furnaces and fireplaces to consume their own smoke. It also provided for sanctions against factories that emitted large amounts of black smoke. The provisions of this law were extended in 1926 with the Smoke Abatement Act to include other emissions, such as soot, ash and gritty particles and to empower local authorities to impose their own regulations.

It was, however, only under the impetus of the Great Smog of 1952 in London, which almost brought the city to a standstill and may have caused upward of 6,000 deaths that the Clean Air Act 1956 was passed and pollution in the city was finally brought to an end. Financial incentives were offered to householders to replace open coal fires with alternatives (such as installing gas fires), or for those who preferred, to burn coke instead (a byproduct of town gas production) which produces minimal smoke. 'Smoke control areas' were introduced in some towns and cities in which only smokeless fuels could be burnt and power stations were relocated away from cities. The act formed an important impetus to modern environmentalism, and caused a rethinking of the dangers of environmental degradation to people's quality of life.

The late 19th century also saw the passage of the first wildlife conservation laws. The zoologist Alfred Newton published a series of investigations into the *Desirability of establishing a 'Close-time' for the preservation of indigenous animals* between 1872 and 1903. His advocacy for legislation to protect animals from hunting during the mating season led to the formation of the Royal Society for the Protection of Birds and influenced the passage of the Sea Birds Preservation Act in 1869 as the first nature protection law in the world.

First Environmental Movements

Early interest in the environment was a feature of the Romantic movement in the early 19th century. The poet William Wordsworth travelled extensively in the Lake District and wrote that it is a "sort of national property in which every man has a right and interest who has an eye to perceive and a heart to enjoy".

John Ruskin an influential thinker who articulated the Romantic ideal of environmental protection and conservation.

Systematic efforts on behalf of the environment only began in the late 19th century; it grew out of the amenity movement in Britain in the 1870s, which was a reaction to industrialization, the growth of cities, and worsening air and water pollution. Starting with the formation of the Commons Preservation Society in 1865, the movement championed rural preservation against the encroachments of industrialisation. Robert Hunter, solicitor for the society, worked with Hardwicke Rawnsley, Octavia Hill, and John Ruskin to lead a successful campaign to prevent the construction of railways to carry slate from the quarries, which would have ruined the unspoilt valleys of Newlands and Ennerdale. This success led to the formation of the Lake District Defence Society (later to become The Friends of the Lake District).

In 1893 Hill, Hunter and Rawnsley agreed to set up a national body to coordinate environmental conservation efforts across the country; the "National Trust for Places of Historic Interest or Natural Beauty" was formally inaugurated in 1894. The organisation obtained secure footing through the 1907 National Trust Bill, which gave the trust the status of a statutory corporation. and the bill was passed in August 1907.

An early "Back-to-Nature" movement, which anticipated the romantic ideal of modern environmentalism, was advocated by intellectuals such as John Ruskin, William Morris, George Bernard Shaw and Edward Carpenter, who were all against consumerism, pollution and other activities that were harmful to the natural world. The movement was a reaction to the urban conditions of the industrial towns, where sanitation was awful, pollution levels intolerable and housing terribly cramped. Idealists championed the rural life as a mythical Utopia and advocated a return to it. John Ruskin argued that people should return to a *small piece of English ground, beautiful, peaceful, and fruitful. We will have no steam engines upon it . . . we will have plenty of flowers and vegetables . . . we will have some music and poetry; the children will learn to dance to it and sing it.*

Practical ventures in the establishment of small cooperative farms were even attempted and old rural traditions, without the "taint of manufacture or the canker of artificiality", were enthusiastically revived, including the Morris dance and the maypole.

These ideas also inspired various environmental groups in the UK, such as the Royal Society for the Protection of Birds, established in 1889 by Emily Williamson as a protest group to campaign for greater protection for the indigenous birds of the island. The Society attracted growing support from the suburban middle-classes as well as support from many other influential figures, such as the ornithologist Professor Alfred Newton. By 1900, public support for the organisation had grown, and it had over 25,000 members. The Garden city movement incorporated many environmental concerns into its urban planning manifesto; the Socialist League and The Clarion movement also began to advocate measures of nature conservation.

WALDEN;

OR,

LIFE IN THE WOODS.

By HENRY D. THOREAU,

AUTHOR OF "A WEEK ON THE CONCORD AND MERRIMACK RIVERS."

I do not propose to write an ode to dejection, but to brag as lustily as chanticleer in the morning, standing on his roost, if only to wake my neighbors up. — Page 92.

BOSTON:

TICKNOR AND FIELDS.

M DCCC LIV.

Original title page of *Walden* by Henry David Thoreau.

The movement in the United States began in the late 19th century, out of concerns for protecting the natural resources of the West, with individuals such as John Muir and Henry David Thoreau making key philosophical contributions. Thoreau was interested in peoples' relationship with nature and studied this by living close to nature in a simple life. He published his experiences in the book *Walden*, which argues that people should become intimately close with nature. Muir came to believe in nature's inherent right, especially after spending time hiking in Yosemite Valley and studying both the ecology and geology. He successfully lobbied congress to form Yosemite National Park and went on to set up the Sierra Club in 1892. The conservationist principles as well as the belief in an inherent right of nature were to become the bedrock of modern environmentalism.

In the 20th century, environmental ideas continued to grow in popularity and recognition. Efforts were starting to be made to save some wildlife, particularly the American bison. The death of the last passenger pigeon as well as the endangerment of the American bison helped to focus the minds of conservationists and popularize their concerns. In 1916 the National Park Service was founded by US President Woodrow Wilson.

The Forestry Commission was set up in 1919 in Britain to increase the amount of woodland in Britain by buying land for afforestation and reforestation. The commission was also tasked with promoting forestry and the production of timber for trade. During the 1920s the Commission focused on acquiring land to begin planting out new forests; much of the land was previously used for agricultural purposes. By 1939 the Forestry Commission was the largest landowner in Britain.

During the 1930s the Nazis had elements that were supportive of animal rights, zoos and wildlife, and took several measures to ensure their protection. In 1933 the government created a stringent animal-protection law and in 1934, *Das Reichsjagdgesetz* (The Reich Hunting Law) was enacted which limited hunting. Several Nazis were environmentalists (notably Rudolf Hess), and species protection and animal welfare were significant issues in the regime. In 1935, the regime enacted the "Reich Nature Protection Act" (*Reichsnaturschutzgesetz*). The concept of the *Dauerwald* (best translated as the "perpetual forest") which included concepts such as forest management and protection was promoted and efforts were also made to curb air pollution.

In 1949, *A Sand County Almanac* by Aldo Leopold was published. It explained Leopold's belief that humankind should have moral respect for the environment and that it is unethical to harm it. The book is sometimes called the most influential book on conservation.

Throughout the 1950s, 1960s, 1970s and beyond, photography was used to enhance public awareness of the need for protecting land and recruiting members to environmental organizations. David Brower, Ansel Adams and Nancy Newhall created the Sierra Club Exhibit Format Series, which helped raise public environmental awareness and brought a rapidly increasing flood of new members to the Sierra Club and to the environmental movement in general. "This Is Dinosaur" edited by Wallace Stegner with photographs by Martin Litton and Philip Hyde prevented the building of dams within Dinosaur National Monument by becoming part of a new kind of activism called environmentalism that combined the conservationist ideals of Thoreau, Leopold and Muir with hard-hitting advertising, lobbying, book distribution, letter writing campaigns, and more. The powerful use of photography in addition to the written word for conservation dated back to the creation of Yosemite National Park, when photographs persuaded Abraham Lincoln to preserve the beautiful glacier carved landscape for all time. The Sierra Club Exhibit Format Series galvanized public opposition to building dams in the Grand Canyon and protected many other national treasures. The Sierra Club often led a coalition of many environmental groups including the Wilderness Society and many others. After a focus on preserving wilderness in the 1950s and 1960s, the Sierra Club and other groups broadened their focus to include such issues as air and water pollution, population concern, and curbing the exploitation of natural resources.

Post-war Expansion

Silent Spring by Rachael Carson, published in 1962, included an endorsement by William O. Douglas.

In 1962, *Silent Spring* by American biologist Rachel Carson was published. The book cataloged the environmental impacts of the indiscriminate spraying of DDT in the US and questioned the logic of releasing large amounts of chemicals into the environment without fully understanding their effects on ecology or human health. The book suggested that DDT and other pesticides may cause cancer and that their agricultural use was a threat to wildlife, particularly birds. The resulting public concern led to the creation of the United States Environmental Protection Agency in 1970 which subsequently banned the agricultural use of DDT in the US in 1972. The limited use of DDT in disease vector control continues to this day in certain parts of the world and remains controversial. The book's legacy was to produce a far greater awareness of environmental issues and interest into how people affect the environment. With this new interest in environment came interest in problems such as air pollution and petroleum spills, and environmental interest grew. New pressure groups formed, notably Greenpeace and Friends of the Earth (US), as well as notable local organizations such as the Wyoming Outdoor Council, which was founded in 1967.

In the 1970s, the environmental movement gained rapid speed around the world as a productive outgrowth of the counterculture movement.

The world's first political parties to campaign on a predominantly environmental platform were the United Tasmania Group Tasmania, Australia and the Values Party of New Zealand. The first green party in Europe was the Popular Movement for the Environment, founded in 1972 in the Swiss canton of Neuchâtel. The first national green party in Europe was PEOPLE, founded in Britain in February 1973, which eventually turned into the Ecology Party, and then the Green Party.

Protection of the environment also became important in the developing world; the Chipko movement was formed in India under the influence of Mohandas Gandhi and they set up peaceful resistance to deforestation by literally hugging trees (leading to the term "tree huggers"). Their peaceful methods of protest and slogan "ecology is permanent economy" were very influential.

Another milestone in the movement was the creation of an Earth Day. Earth Day was first observed in San Francisco and other cities on March 21, 1970, the first day of spring. It was created to give awareness to environmental issues. On March 21, 1971, United Nations Secretary-General U Thant spoke of a spaceship Earth on Earth Day, hereby referring to the ecosystem services the earth supplies to us, and hence our obligation to protect it (and with it, ourselves). Earth Day is now coordinated globally by the Earth Day Network, and is celebrated in more than 175 countries every year.

The UN's first major conference on international environmental issues, the United Nations Conference on the Human Environment (also known as the Stockholm Conference), was held on June 5–16, 1972. It marked a turning point in the development of international environmental politics.

By the mid-1970s, many felt that people were on the edge of environmental catastrophe. The Back-to-the-land movement started to form and ideas of environmental ethics joined with anti-Vietnam War sentiments and other political issues. These individuals lived outside normal society and started to take on some of the more radical environmental theories such as deep ecology. Around this time more mainstream environmentalism was starting to show force with the signing of the Endangered Species Act in 1973 and the formation of CITES in 1975. Significant amendments were also enacted to the United States Clean Air Act and Clean Water Act.

In 1979, James Lovelock, a British scientist, published *Gaia: A new look at life on Earth*, which put forth the Gaia hypothesis; it proposes that life on earth can be understood as a single organism. This became an important part of the Deep Green ideology. Throughout the rest of the history of environmentalism there has been debate and argument between more radical followers of this Deep Green ideology and more mainstream environmentalists.

Today

Environmentalism continues to evolve to face up to new issues such as global warming, overpopulation and genetic engineering.

Recent research demonstrates a precipitous decline in the public's interest in 19 different areas of environmental concern. Americans are less likely be actively participating in an environmental movement or organization and more likely to identify as "unsympathetic" to an environmental movement then in 2000. This is likely a lingering factor of the Great Recession in 2008. Since 2005 the percentage of Americans agreeing that the environment should be given priority over economic growth has dropped 10 points, in contrast, those feeling that growth should be given priority "even if the environment suffers to some extent" has risen 12 percent. These numbers point to the growing complexity of environmentalism and its relationship to economics.

Environmental Movement

The *environmental movement* (a term that sometimes includes the conservation and green movements) is a diverse scientific, social, and political movement. Though the movement is represented by a range of organizations, because of the inclusion of environmentalism in the classroom curriculum, the environmental movement has a younger demographic than is common in other social movements.

Before flue-gas desulfurization was installed, the air-polluting emissions from this power plant in New Mexico contained excessive amounts of sulfur dioxide.

Environmentalism as a movement covers broad areas of institutional oppression, including for example: consumption of ecosystems and natural resources into waste, dumping waste into disadvantaged communities, air pollution, water pollution, weak infrastructure, exposure of organic life to toxins, mono-culture, anti-polythene drive (jhola movement) and various other focuses. Because of these divisions, the environmental movement can be categorized into these primary focuses: environmental science, environmental activism, environmental advocacy, and environmental justice.

Free Market Environmentalism

Free market environmentalism is a theory that argues that the free market, property rights, and tort law provide the best tools to preserve the health and sustainability of the environment. It considers environmental stewardship to be natural, as well as the expulsion of polluters and other aggressors through individual and class action. It has been supported by libertarians and many conservatives.

Evangelical Environmentalism

Evangelical environmentalism is an environmental movement in the United States in which some Evangelicals have emphasized biblical mandates concerning humanity's role as steward and subsequent responsibility for the caretaking of Creation. While the movement has focused on different environmental issues, it is best known for its focus of addressing climate action from a biblically grounded theological perspective. The Evangelical Climate Initiative argues that human-induced climate change will have severe consequences and impact the poor the hardest, and that God's mandate to Adam to care for the Garden of Eden also applies to evangelicals today, and that it is therefore a moral obligation to work to mitigate climate impacts and support communities in adapting to change.

Preservation and Conservation

Environmental preservation in the United States and other parts of the world, including Australia, is viewed as the setting aside of natural resources to prevent damage caused by contact with humans or by certain human activities, such as logging, mining, hunting, and fishing, often to replace them with new human activities such as tourism and recreation. Regulations and laws may be enacted for the preservation of natural resources.

Organizations and Conferences

Reef doctor work station in Ifaty, Madagascar.

Environmental organizations can be global, regional, national or local; they can be government-run or private (NGO). Environmentalist activity exists in almost every country. Moreover, groups dedicated to community development and social justice also focus on environmental concerns.

There are some volunteer organizations. For example, Ecoworld and Paryawaran Sachetak Samiti are environmental organizations which are based on teamwork and volunteer work. Some US environmental organizations, among them the Natural Resources Defense Council and the Environmental Defense Fund, specialize in bringing lawsuits (a tactic seen as particularly useful in that country). Other groups, such as the US-based National Wildlife Federation, the Nature Conservancy, and The Wilderness Society, and global groups like the World Wide Fund for Nature and Friends of the Earth, disseminate information, participate in public hearings, lobby, stage demonstrations, and may purchase land for preservation. Statewide nonprofit organizations such as the Wyoming Outdoor Council often collaborate with these national organizations and employ similar strategies. Smaller groups, including Wildlife Conservation International, conduct research on endangered species and ecosystems. More radical organizations, such as Greenpeace, Earth First!, and the Earth Liberation Front, have more directly opposed actions they regard as environmentally harmful. While Greenpeace is devoted to nonviolent confrontation as a means of bearing witness to environmental wrongs and bringing issues into the public realm for debate, the underground *Earth Liberation Front* engages in the clandestine destruction of property, the release of caged or penned animals, and other criminal acts. Such tactics are regarded as unusual within the movement, however.

On an international level, concern for the environment was the subject of a United Nations Conference on the Human Environment in Stockholm in 1972, attended by 114 nations. Out of this meeting developed UNEP (United Nations Environment Programme) and the follow-up United Nations Conference on Environment and Development in 1992. Other international organizations in support of environmental policies development include the Commission for Environmental Cooperation (as part of NAFTA), the European Environment Agency (EEA), and the Intergovernmental Panel on Climate Change (IPCC).

Environmental Protests

Climate activists blockade British Airports Authority's headquarters for day of action.

"March Against Monsanto", Vancouver, Canada, May 25, 2013.

Notable environmental protests and campaigns include:

- 2010 Xinfa aluminum plant protest.
- Anti-WAAhnsinns Festival.
- Camp for Climate Action.
- Campaign against Climate Change.
- Climate Rush.
- Cofán people oil drilling protest (Ecuador).
- Earth Day.
- Earth First!
- Earthlife Africa.
- Global Day of Action.
- Gurindji Strike.
- Hands off our Forest.
- Homes before Roads.
- Kupa Piti Kungka Tjuta.
- Love Canal protests.
- March Against Monsanto.
- Nevada Desert Experience.
- Plane Mad.
- Plane Stupid.
- Qidong protest.
- Save Manapouri Campaign.
- Say Yes demonstrations.
- Shifang protest.
- Stop Climate Chaos.

Environmentalists

Notable advocates for environmental protection and sustainability include:

- Edward Abbey (author).
- David Attenborough (broadcaster, naturalist).
- John James Audubon (naturalist).

- David Bellamy (botanist).
- Wendell Berry (farmer, philosopher).
- David Brower (writer, activist).
- Lester Brown (environmental analyst, author).
- Kevin Buzzacott (Aboriginal activist).
- Helen Caldicott (medical doctor).
- Rachel Carson (biologist, writer)
- Prince Charles (British Royal Family member).
- Barry Commoner (biologist, politician).
- Jacques-Yves Cousteau (explorer, ecologist).
- Peter Dauvergne (political scientist).
- Leonardo Dicaprio (actor and environmentalist).
- Paul R. Ehrlich (population biologist).
- Hans-Josef Fell (German Green Party member).
- Jane Fonda (actor).
- Mizuho Fukushima (politician, activist).
- Peter Garrett (musician, politician).
- Al Gore (former Vice President of the United States).
- James Hansen (scientist).
- Garrett Hardin (ecologist, ecophilosopher).
- Denis Hayes (environmentalist and solar power advocate).
- Tetsunari Iida (sustainable energy advocate).
- Aldo Leopold (ecologist).
- A. Carl Leopold (plant physiologist).
- James Lovelock (scientist).
- Amory Lovins (energy policy analyst).
- Hunter Lovins (environmentalist).
- Caroline Lucas (politician).
- Bill McKibben (writer, activist).
- David McTaggart (activist).

- Chico Mendes (activist).
- George Monbiot (journalist).
- John Muir (naturalist, activist).
- Ralph Nader (activist).
- Gaylord Nelson (politician).
- Alan Pears (environmental consultant and energy efficiency pioneer).
- Gifford Pinchot (first chief of the USFS).
- Jonathon Porritt (politician).
- John Wesley Powell (second director of the USGS).
- Barbara Pyle (documentarian and executive producer of *Captain Planet and the Planeteers*).
- Phil Radford (environmental, clean energy and democracy advocate, Greenpeace Executive Director).
- Bonnie Raitt (musician).
- Theodore Roosevelt (former President of the United States).
- E. F. Schumacher (author of *Small is Beautiful*).
- Swami Sundaranand {Yogi, photographer, and mountaineer}.
- Cass Sunstein (environmental lawyer).
- David Suzuki (scientist, broadcaster).
- Henry David Thoreau (writer, philosopher).
- Stewart Udall (former United States Secretary of the Interior).
- Jo Valentine (politician and activist).
- Dominique Voynet (politician and environmentalist).
- Gabriel Willow (environmental educator, naturalist).
- Howard Zahniser (author of the 1964 Wilderness Act).

Assassination of Environmental Activists

Every year, more than 100 environmentalist activists are assassinated throughout the world.

Usage in Popular Culture

- The popular media have been used to convey conservation messages in the U.S. For instance, the U.S. Forest Service created Smokey the Bear in 1944; he appeared in countless posters, radio and television programs, movies, press releases, and other guises to warn about forest fires. The comic strip Mark Trail, by environmentalist Ed Dodd, began in 1946; it still appears weekly in 175 newspapers. Another example is the children's animated show *Captain Planet and the Planeteers*, created by Ted Turner and Barbara Pyle in 1989 to inform kids on environmental issues. The show aired for six seasons and 113 episodes, in 100 countries worldwide from 1990 to 1996.

- Miss Earth is one of the three largest international beauty pageants alongside Miss Universe and Miss World that promotes Environmental Awareness. The reigning titleholders dedicate their year to promote specific projects and often address issues concerning the environment and other global issues through school tours, tree planting activities, street campaigns, coastal clean ups, speaking engagements, shopping mall tours, media guesting, environmental fair, storytelling programs, eco-fashion shows, and other environmental activities. The Miss Earth winner is the spokesperson for the Miss Earth Foundation, the United Nations Environment Programme (UNEP) and other environmental organizations. The Miss Earth Foundation also works with the environmental departments and ministries of participating countries, various private sectors and corporations, as well as Greenpeace and the World Wildlife Foundation (WWF).

An Alternative View

Many environmentalists believe that human interference with 'nature' should be restricted or minimised as a matter of urgency (for the sake of life, or the planet, or just for the benefit of the human species), whereas environmental skeptics and anti-environmentalists do not believe that there is such a need. One can also regard oneself as an environmentalist and believe that human 'interference' with 'nature' should be *increased*. Nevertheless, there is a risk that the shift from emotional environmentalism into the technical management of natural resources and hazards could decrease the touch of humans with nature, leading to less concern with environment preservation.

References

- Cockell, Charles; Koeberl, Christian; Gilmour, Iain (18 May 2006). Biological Processes Associated with Impact Events (1 ed.). Springer Science & Business Media. pp. 197–219. ISBN 978-3-540-25736-3
- Lewis, E.E.; Campbell, J.F.; Sukhdeo, M.V.K. (2002). The Behavioural Ecology of Parasites. CABI. p. 183. ISBN 978-0-85199-754-4
- Abe, Y.; Bignell, David Edward; Higashi, T. (2014). Termites: Evolution, Sociality, Symbioses, Ecology. Springer. p. 437. ISBN 978-94-017-3223-9
- John G. Kelcey, John G. (2015). Vertebrates and Invertebrates of European Cities:Selected Non-Avian Fauna. Springer. p. 124. ISBN 978-1-4939-1698-6
- Cole, Bob (March 2008). "Appendix 6". The SAA BUhlmann DeeP-Stop System Handbook. Sub-Aqua Association. pp. vi–1. ISBN 978-0-9532904-8-2
- Miller, G. Tyler; Spoolman, Scott (2008). Living in the Environment: Principles, Connections, and Solutions. Cengage Learning. pp. 193–195. ISBN 978-0-495-55671-8
- G. Miller; Scott Spoolman (2012). Environmental Science - Biodiversity Is a Crucial Part of the Earth's Natural Capital. Cengage Learning. p. 62. ISBN 1-133-70787-4. Retrieved 16, February 2020
- Hamilton Raven, Peter; Brooks Johnson, George (2002). Biology. McGraw-Hill Education. p. 68. ISBN 978-0-07-112261-0. Retrieved 05, April 2020

- Tor-Björn Larsson (2001). Biodiversity evaluation tools for European forests. Wiley-Blackwell. p. 178. ISBN 978-87-16-16434-6. Retrieved 29, January 2020
- Davis. Intro To Env Engg (Sie), 4E. McGraw-Hill Education (India) Pvt Ltd. pp. 4–. ISBN 978-0-07-067117-1. Retrieved 18, June 2020
- Sala, Osvaldo E.; Meyerson, Laura A.; Parmesan, Camille (26 January 2009). Biodiversity change and human health: from ecosystem services to spread of disease. Island Press. pp. 3–5. ISBN 978-1-59726-497-6. Retrieved 13, May 2020
- "Petroleum fly". Grzimek's Animal Life Encyclopedia. Volume 3: Insects (2nd ed.). The Gale Group. 2004. p. 367. ISBN 0-7876-5779-4
- Lindenmayer, David B.; Fischer, Joern (2013). Habitat Fragmentation and Landscape Change: An Ecological and Conservation Synthesis. Island Press. pp. 1–10. ISBN 978-1-59726-606-2

Chapter 5

Biogeochemical Cycles and its Types

Biogeochemical cycle refers to any of the natural pathways by which essential elements of living matter are circulated through the various biotic and abiotic parts of an ecosystem. Water cycle, carbon cycle, nitrogen cycle, oxygen cycle, phosphorus cycle, sulfur cycle, etc. are all part of the biogeochemical cycle. This chapter deals with the various biogeochemical cycles in a comprehensive manner for the benefit of the reader.

Biogeochemical Cycle

A biogeochemical cycle is one of several natural cycles, in which conserved matter moves through the biotic and abiotic parts of an ecosystem.

In biology, conserved matter refers to the finite amount of matter, in the form of atoms, that is present within the Earth. Since, according to the Law of Conservation of Mass, matter cannot be created or destroyed, all atoms of matter are cycled through Earth's systems albeit in various forms.

In other words, the Earth only receives energy from the sun, which is given off as heat, whilst all other chemical elements remain within a closed system.

The main chemical elements that are cycled are: carbon (C), hydrogen (H), nitrogen (N), oxygen (O), phosphorous (P) and sulfur (S). These are the building blocks of life, and are used for essential processes, such as metabolism, the formation of amino acids, cell respiration and the building of tissues.

These fundamental elements can be easily remembered with the acronym CHNOPS.

Each of these elements is circulated through the biotic components, which are the living parts of an ecosystem, and the abiotic components, which are the non-living parts.

The abiotic components can be subdivided into three categories: thehydrosphere (water), the atmosphere (air) and the lithosphere(rock).

The biosphere is a term which can be used to describe the system that contains all living organisms, including plants, animals and bacteria, as well as their interactions among and between each other, and their interactions with the Earth's abiotic systems. The biosphere is sometimes called the ecosphere, and can be defined as the sum of all ecosystems.

With this knowledge, the words "biogeochemical cycle" can be easily broken down. "Bio-" is the biotic system, "geo-" is the geological component, and "chemical" is the elements which are moved through a "cycle".

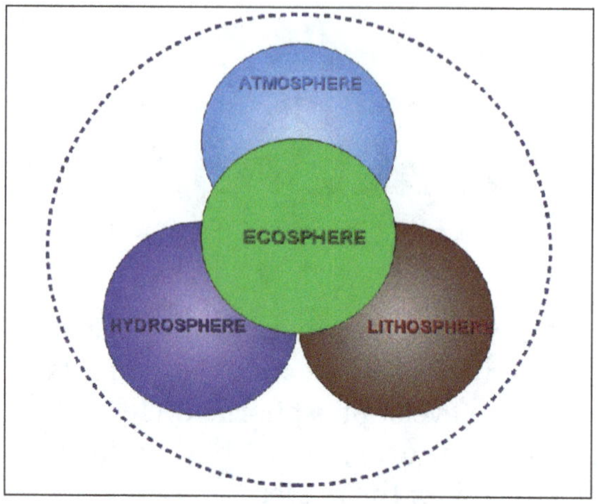

Biosphere system.

At particular stages of their cycling, any of the elements may be stored and accumulated within a particular place for a long period time (e.g. within a rocky substrate, or in the atmosphere). These places are called "sinks" or "reservoirs".

A "source" is anything from which an element is output, for example volcanoes give off large amounts of carbon in the form of CO_2, while human waste is a source for nitrogen, sulfur and phosphorous.

Systems

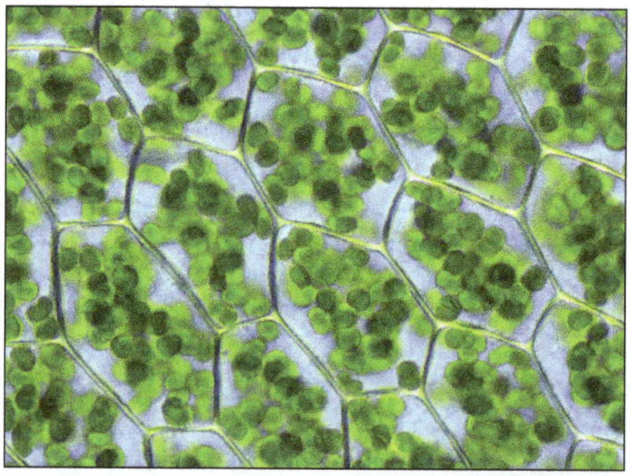

Chloroplasts conduct photosynthesis and are found in plant cells and other eukaryoticorganisms. These are Chloroplasts visible in the cells of Plagiomnium affine — Many-fruited Thyme-moss.

Ecological systems (ecosystems) have many biogeochemical cycles operating as a part of the system, for example the water cycle, the carbon cycle, the nitrogen cycle, etc. All chemical elements occurring in organisms are part of biogeochemical cycles. In addition to being a part of living organisms, these chemical elements also cycle through abiotic factors of ecosystems such as water (hydrosphere), land (lithosphere), and/or the air (atmosphere).

The living factors of the planet can be referred to collectively as the biosphere. All the nutrients—such as carbon, nitrogen, oxygen, phosphorus, and sulfur—used in ecosystems by living organisms are a part of a closed system; therefore, these chemicals are recycled instead of being lost and replenished constantly such as in an open system.

The flow of energy in an ecosystem is an open system; the sun constantly gives the planet energy in the form of light while it is eventually used and lost in the form of heat throughout the trophic levels of a food web. Carbon is used to make carbohydrates, fats, and proteins, the major sources of food energy. These compounds are oxidized to release carbon dioxide, which can be captured by plants to make organic compounds. The chemical reaction is powered by the light energy of the sun.

It is possible for an ecosystem to obtain energy without sunlight. Carbon must be combined with hydrogen and oxygen in order to be utilized as an energy source, and this process depends on sunlight. Ecosystems in the deep sea, where no sunlight can penetrate, use sulfur. Hydrogen sulfide near hydrothermal vents can be utilized by organisms such as the giant tube worm. In the sulfur cycle, sulfur can be forever recycled as a source of energy. Energy can be released through the oxidation and reduction of sulfur compounds (e.g., oxidizing elemental sulfur to sulfite and then to sulfate).

Although the Earth constantly receives energy from the sun, its chemical composition is essentially fixed, as additional matter is only occasionally added by meteorites. Because this chemical composition is not replenished like energy, all processes that depend on these chemicals must be recycled. These cycles include both the living biosphere and the nonliving lithosphere, atmosphere, and hydrosphere.

Reservoirs

Coal is a reservoir of carbon.

The chemicals are sometimes held for long periods of time in one place. This place is called a reservoir, which, for example, includes such things as coaldeposits that are storing carbon for a long period of time. When chemicals are held for only short periods of time, they are being held in exchange pools. Examples of exchange pools include plants and animals.

Plants and animals utilize carbon to produce carbohydrates, fats, and proteins, which can then be used to build their internal structures or to obtain energy. Plants and animals temporarily use carbon in their systems and then release it back into the air or surrounding medium. Generally, reservoirs are abiotic factors whereas exchange pools are biotic factors. Carbon is held for a relatively short time in plants and animals in comparison to coal deposits. The amount of time that a chemical is held in one place is called its residence.

Important Cycles

The most well-known and important biogeochemical cycles are shown below:

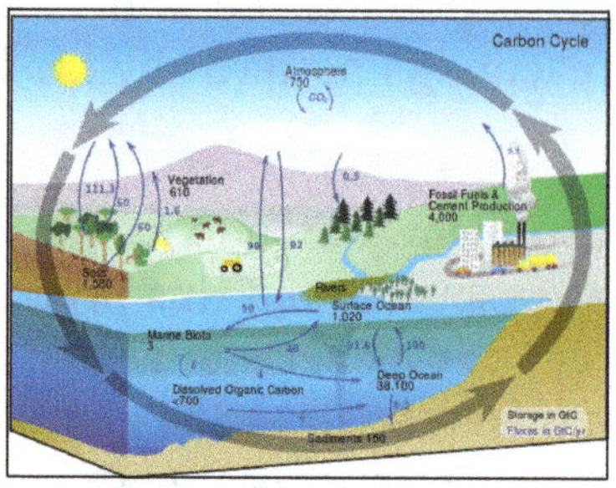

Carbon cycle.

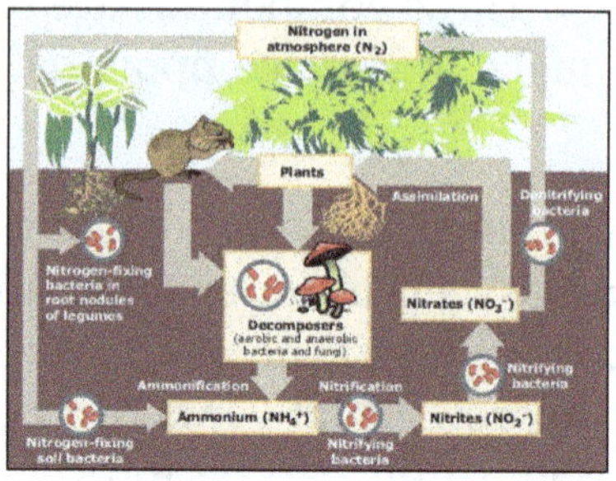

Nitrogen cycle.

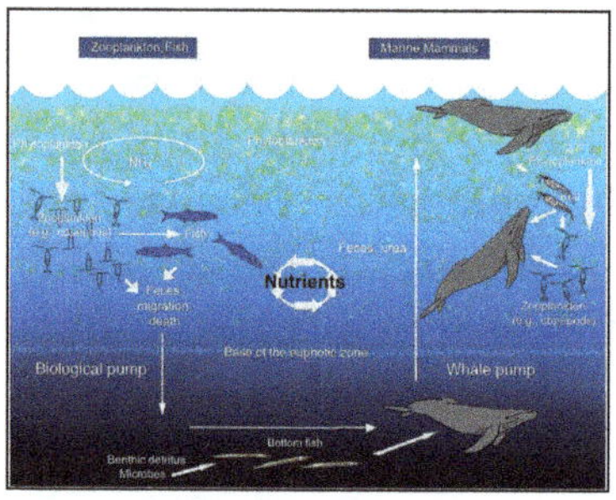

Nutrient cycle.

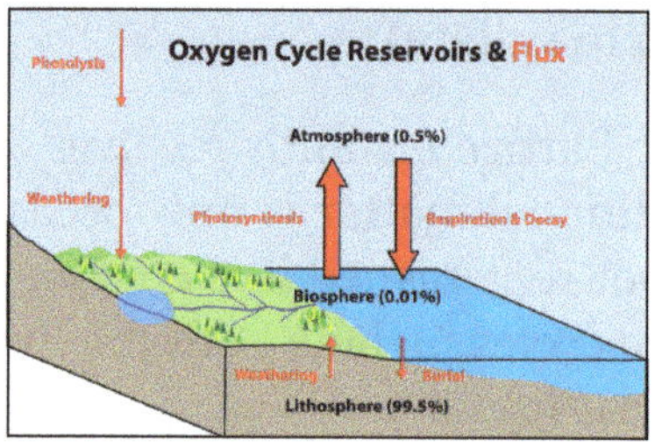

Oxygen cycle.

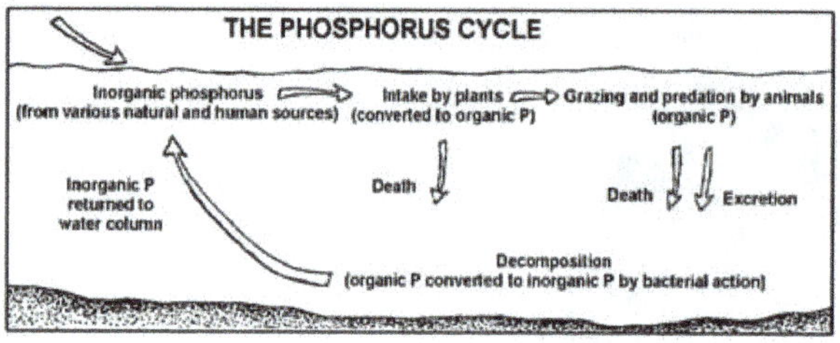

Phosphorus cycle.

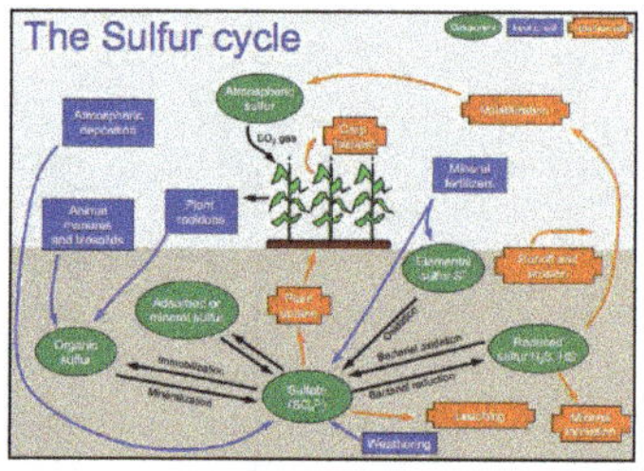

Sulfur cycle.

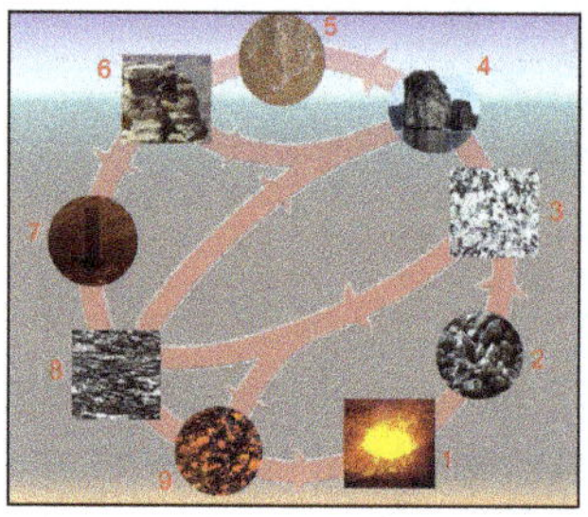

Rock cycle.

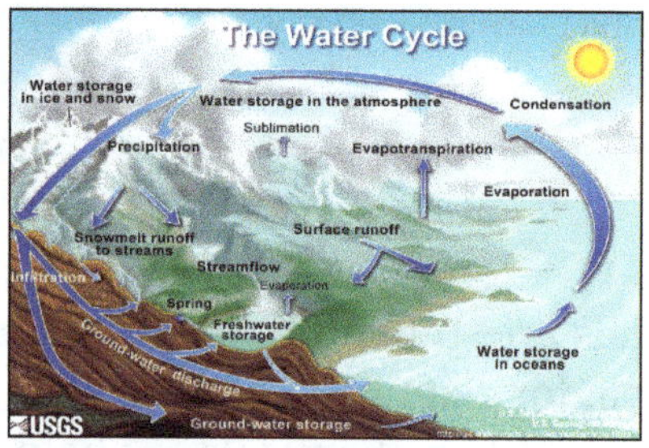

Water cycle.

Water Cycle

Water cycle, also called hydrologic cycle, cycle that involves the continuous circulation of water in the Earth-atmosphere system. Of the many processes involved in the water cycle, the most important are evaporation, transpiration, condensation, precipitation, and runoff. Although the total amount of water within the cycle remains essentially constant, its distribution among the various processes is continually changing.

Evaporation, one of the major processes in the cycle, is the transfer of water from the surface of the Earth to the atmosphere. By evaporation, water in the liquid state is transferred to the gaseous, or vapour, state. This transfer occurs when some molecules in a water mass have attained sufficient kinetic energy to eject themselves from the water surface. The main factors affecting evaporation are temperature, humidity, wind speed, and solar radiation. The direct measurement of evaporation, though desirable, is difficult and possible only at point locations. The principal source of water vapour is the oceans, but evaporation also occurs in soils, snow, and ice. Evaporation from snow and ice, the direct conversion from solid to vapour, is known as sublimation. Transpiration is the evaporation of water through minute pores, or stomata, in the leaves of plants. For practical purposes, transpiration and the evaporation from all water, soils, snow, ice, vegetation, and other surfaces are lumped together and called evapotranspiration, or total evaporation.

The water cycle begins and ends in the ocean.Created and produced by QA International.

Water vapour is the primary form of atmospheric moisture. Although its storage in the atmosphere is comparatively small, water vapour is extremely important in forming the moisture supply for dew, frost, fog, clouds, and precipitation. Practically all water vapour in the atmosphere is confined to the troposphere (the region below 6 to 8 miles [10 to 13 km] altitude).

The transition process from the vapour state to the liquid state is called condensation. Condensation may take place as soon as the air contains more water vapour than it can receive from a free water surface through evaporation at the prevailing temperature. This condition occurs as the consequence of either cooling or the mixing of air masses of different temperatures. By condensation, water vapour in the atmosphere is released to form precipitation.

Precipitation that falls to the Earth is distributed in four main ways: some is returned to the atmosphere by evaporation, some may be intercepted by vegetation and then evaporated from the surface of leaves, some percolates into the soil by infiltration, and the remainder flows directly as surface runoff into the sea. Some of the infiltrated precipitation may later percolate into streams as groundwater runoff. Direct measurement of runoff is made by stream gauges and plotted against time on hydrographs.

Most groundwater is derived from precipitation that has percolated through the soil. Groundwater flow rates, compared with those of surface water, are very slow and variable, ranging from a few millimetres to a few metres a day. Groundwater movement is studied by tracer techniques and remote sensing.

Ice also plays a role in the water cycle. Ice and snow on the Earth's surface occur in various forms such as frost, sea ice, and glacier ice. When soil moisture freezes, ice also occurs beneath the Earth's surface, forming permafrost in tundra climates. About 18,000 years ago glaciers and ice caps covered approximately one-third of the Earth's land surface. Today about 12 percent of the land surface remains covered by ice masses.

Description

The sun, which drives the water cycle, heats water in oceans and seas. Water evaporates as water vapor into the air. Some ice and snow sublimates directly into water vapor. Evapotranspiration is water transpired from plants and evaporated from the soil. The water molecule H_2O has smaller molecular mass than the major components of the atmosphere, nitrogen and oxygen, N_2 and O_2, hence is less dense. Due to the significant difference in density, buoyancy drives humid air higher. As altitude increases, air pressure decreases and the temperature drops. The lower temperature causes water vapor to condense into tiny liquid water droplets which are heavier than the air, and fall unless supported by an updraft. A huge concentration of these droplets over a large space up in the atmosphere become visible as cloud. Some condensation is near ground level, and called fog.

Atmospheric circulation moves water vapor around the globe, cloud particles collide, grow, and fall out of the upper atmospheric layers as precipitation. Some precipitation falls as snow or hail, sleet, and can accumulate as ice caps and glaciers, which can store frozen water for thousands of years. Most water falls back into the oceans or onto land as rain, where the water flows over the ground as surface runoff. A portion of runoff enters rivers in valleys in the landscape, with streamflow moving water towards the oceans. Runoff and water emerging from the ground (groundwater) may be stored as freshwater in lakes. Not all runoff flows into rivers, much of it soaks into the ground as infiltration. Some water infiltrates deep into the ground and replenishes aquifers, which can store freshwater for long periods of time. Some infiltration stays close to the land surface and can seep back into surface-water bodies (and the ocean) as groundwater discharge. Some groundwater finds openings in the land surface and comes out as freshwater springs. In river valleys and floodplains, there is often continuous water exchange between surface water and ground water in the hyporheic zone. Over time, the water returns to the ocean, to continue the water cycle.

Processes

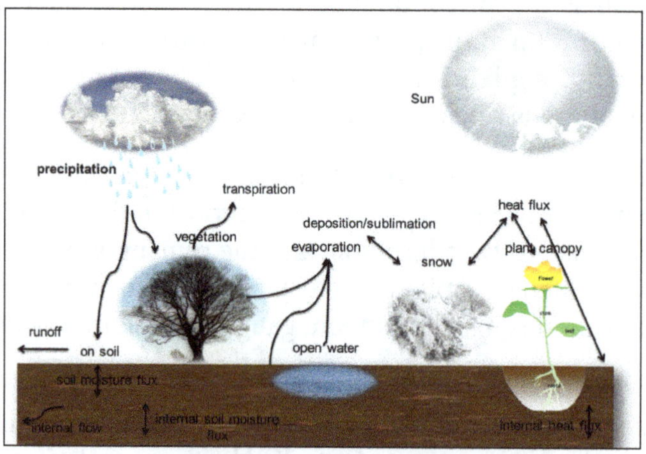

Many different processes lead to movements and phase changes in water.

Precipitation

Condensed water vapor that falls to the Earth's surface. Most precipitation occurs as rain, but also includes snow, hail, fog drip, graupel, and sleet. Approximately 505,000 km³ (121,000 cu mi) of water falls as precipitation each year, 398,000 km³ (95,000 cu mi) of it over the oceans. The rain on land contains 107,000 km³ (26,000 cu mi) of water per year and a snowing only 1,000 km³ (240 cu mi). 78% of global precipitation occurs over the ocean.

Canopy interception

The precipitation that is intercepted by plant foliage eventually evaporates back to the atmosphere rather than falling to the ground.

Snowmelt

The runoff produced by melting snow.

Runoff

The variety of ways by which water moves across the land. This includes both surface runoff and channel runoff. As it flows, the water may seep into the ground, evaporate into the air, become stored in lakes or reservoirs, or be extracted for agricultural or other human uses.

Infiltration

The flow of water from the ground surface into the ground. Once infiltrated, the water becomes soil moisture or groundwater. A recent global study using water stable isotopes, however, shows that not all soil moisture is equally available for groundwater recharge or for plant transpiration.

Subsurface flow

The flow of water underground, in the vadose zone and aquifers. Subsurface water may return to the surface (e.g. as a spring or by being pumped) or eventually seep into the oceans. Water returns to the land surface at lower elevation than where it infiltrated, under the force of gravity or gravity induced pressures. Groundwater tends to move slowly and is replenished slowly, so it can remain in aquifers for thousands of years.

Evaporation

The transformation of water from liquid to gas phases as it moves from the ground or bodies of water into the overlying atmosphere. The source of energy for evaporation is primarily solar radiation. Evaporation often implicitly includes transpiration from plants, though together they are specifically referred to as evapotranspiration. Total annual evapotranspiration amounts to approximately 505,000 km³ (121,000 cu mi) of water, 434,000 km³(104,000 cu mi) of which evaporates from the oceans. 86% of global evaporation occurs over the ocean.

Sublimation

The state change directly from solid water (snow or ice) to water vapor by passing the liquid state.

Deposition

This refers to changing of water vapor directly to ice.

Advection

The movement of water through the atmosphere. Without advection, water that evaporated over the oceans could not precipitate over land.

Condensation

The transformation of water vapor to liquid water droplets in the air, creating clouds and fog.

Transpiration

The release of water vapor from plants and soil into the air.

Percolation

Water flows vertically through the soil and rocks under the influence of gravity.

Plate tectonics

Water enters the mantle via subduction of oceanic crust. Water returns to the surface via volcanism.

The water cycle involves many of these processes.

Residence Times

The residence time of a reservoir within the hydrologic cycle is the average time a water molecule will spend in that reservoir. It is a measure of the average age of the water in that reservoir.

Average reservoir residence times	
Reservoir	Average residence time
Antarctica	20,000 years
Oceans	3,200 years
Glaciers	20 to 100 years
Seasonal snow cover	2 to 6 months
Soil moisture	1 to 2 months
Groundwater: shallow	100 to 200 years
Groundwater: deep	10,000 years
Lakes (see lake retention time)	50 to 100 years
Rivers	2 to 6 months
Atmosphere	9 days

Groundwater can spend over 10,000 years beneath Earth's surface before leaving. Particularly old groundwater is called fossil water. Water stored in the soil remains there very briefly, because it is spread thinly across the Earth, and is readily lost by evaporation, transpiration, stream flow, or groundwater recharge. After evaporating, the residence time in the atmosphere is about 9 days before condensing and falling to the Earth as precipitation.

The major ice sheets – Antarctica and Greenland – store ice for very long periods. Ice from Antarctica has been reliably dated to 800,000 years before present, though the average residence time is shorter.

In hydrology, residence times can be estimated in two ways. The more common method relies on the principle of conservation of mass and assumes the amount of water in a given reservoir is roughly constant. With this method, residence times are estimated by dividing the volume of the reservoir by the rate by which water either enters or exits the reservoir. Conceptually, this is equivalent to timing how long it would take the reservoir to become filled from empty if no water were to leave (or how long it would take the reservoir to empty from full if no water were to enter).

An alternative method to estimate residence times, which is gaining in popularity for dating groundwater, is the use of isotopic techniques. This is done in the subfield of isotope hydrology.

Changes Over Time

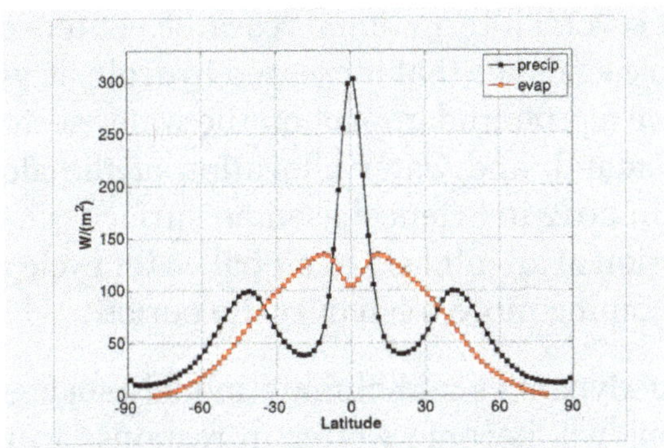

Time-mean precipitation and evaporation as a function of latitude as simulated by an aqua-planet version of an atmospheric GCM (GFDL's AM2.1) with a homogeneous "slab-ocean" lower boundary (saturated surface with small heat capacity), forced by annual mean insolation.

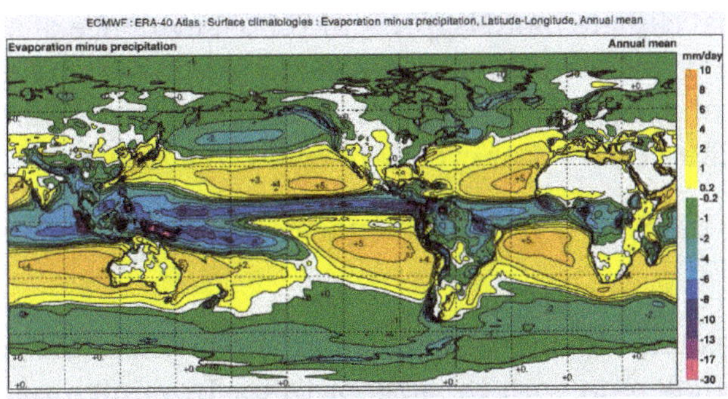

Global map of annual mean evaporation minus precipitation by latitude-longitude.

The water cycle describes the processes that drive the movement of water throughout the hydrosphere. However, much more water is "in storage" for long periods of time than is actually moving through the cycle. The storehouses for the vast majority of all water on Earth are the oceans. It is estimated that of the 332,500,000 mi^3 (1,386,000,000 km^3) of the world's water supply, about 32 1,000,000 mi^3 (1,338,000,000 km^3) is stored in oceans, or about 97%. It is also estimated that the oceans supply about 90% of the evaporated water that goes into the water cycle.

During colder climatic periods, more ice caps and glaciers form, and enough of the global water supply accumulates as ice to lessen the amounts in other parts of the water cycle. The reverse is true during warm periods. During the last ice age, glaciers covered almost one-third of Earth's land mass with the result being that the oceans were about 122 m (400 ft) lower than today. During the last global "warm spell," about 125,000 years ago, the seas were about 5.5 m (18 ft) higher than they are now. About three million years ago the oceans could have been up to 50 m (165 ft) higher.

The scientific consensus expressed in the 2007 Intergovernmental Panel on Climate Change (IPCC) Summary for Policymakers is for the water cycle to continue to intensify throughout the 21st century, though this does not mean that precipitation will increase in all regions. In subtropical land areas — places that are already relatively dry — precipitation is projected to decrease during the 21st century, increasing the probability of drought. The drying is projected to be strongest near the poleward margins of the subtropics (for example, the Mediterranean Basin, South Africa, southern Australia, and the Southwestern United States). Annual precipitation amounts are expected to increase in near-equatorial regions that tend to be wet in the present climate, and also at high latitudes. These large-scale patterns are present in nearly all of the climate model simulations conducted at several international research centers as part of the 4th Assessment of the IPCC. There is now ample evidence that increased hydrologic variability and change in climate has and will continue to have a profound impact on the water sector through the hydrologic cycle, water availability, water demand, and water allocation at the global, regional, basin, and local levels. Research published in 2012 in Sciencebased on surface ocean salinity over the period 1950 to 2000 confirm this projection of an intensified global water cycle with salty areas becoming more saline and fresher areas becoming more fresh over the period:

> Fundamental thermodynamics and climate models suggest that dry regions will become drier and wet regions will become wetter in response to warming. Efforts to detect this long-term response in sparse surface observations of rainfall and evaporation remain ambiguous. We show that ocean salinity patterns express an identifiable fingerprint of an intensifying water cycle. Our 50-year observed global surface salinity changes, combined with changes from global climate models, present robust evidence of an intensified global water cycle at a rate of 8 ± 5% per degree of surface warming. This rate is double the response projected by current-generation climate models and suggests that a substantial (16 to 24%) intensification of the global water cycle will occur in a future 2° to 3° warmer world.

An instrument carried by the SAC-D satellite Aquarius, launched in June, 2011, measured global sea surface salinity.

Glacial retreat is also an example of a changing water cycle, where the supply of water to glaciers from precipitation cannot keep up with the loss of water from melting and sublimation. Glacial retreat since 1850 has been extensive.

Human activities that alter the water cycle include:

- Agriculture.
- Industry.
- Alteration of the chemical composition of the atmosphere.
- Construction of dams.

- Deforestation and afforestation.
- Removal of groundwater from wells.
- Water abstraction from rivers.
- Urbanization.

Effects on Climate

The water cycle is powered from solar energy. 86% of the global evaporation occurs from the oceans, reducing their temperature by evaporative cooling. Without the cooling, the effect of evaporation on the greenhouse effect would lead to a much higher surface temperature of 67 °C (153 °F), and a warmer planet.

Aquifer drawdown or overdrafting and the pumping of fossil water increases the total amount of water in the hydrosphere, and has been postulated to be a contributor to sea-level rise.

Effects on Biogeochemical Cycling

While the water cycle is itself a biogeochemical cycle, flow of water over and beneath the Earth is a key component of the cycling of other biogeochemicals.Runoff is responsible for almost all of the transport of eroded sediment and phosphorus from land to waterbodies. The salinity of the oceans is derived from erosion and transport of dissolved salts from the land. Cultural eutrophication of lakes is primarily due to phosphorus, applied in excess to agricultural fields in fertilizers, and then transported overland and down rivers. Both runoff and groundwater flow play significant roles in transporting nitrogen from the land to waterbodies. The dead zone at the outlet of the Mississippi River is a consequence of nitrates from fertilizer being carried off agricultural fields and funnelled down the river system to the Gulf of Mexico. Runoff also plays a part in the carbon cycle, again through the transport of eroded rock and soil.

Slow Loss over Geologic Time

The hydrodynamic wind within the upper portion of a planet's atmosphere allows light chemical elements such as Hydrogen to move up to the exobase, the lower limit of the exosphere, where the gases can then reach escape velocity, entering outer space without impacting other particles of gas. This type of gas loss from a planet into space is known as planetary wind. Planets with hot lower atmospheres could result in humid upper atmospheres that accelerate the loss of hydrogen.

The water cycle has a dramatic influence on Earth's climate and ecosystems.

Climate is all the weather conditions of an area, evaluated over a period of time. Two weather conditions that contribute to climate include humidity and temperature. These weather conditions are influenced by the water cycle.

Humidity is simply the amount of water vapor in the air. As water vapor is not evenly distributed by the water cycle, some regions experience higher humidity than others. This contributes to radically different climates. Islands or coastal regions, where water vapor makes up more of the atmosphere, are usually much more humid than inland regions, where water vapor is scarcer.

A region's temperature also relies on the water cycle. Through the water cycle, heat is exchanged and temperatures fluctuate. As water evaporates, for example, it absorbs energy and cools the local environment. As water condenses, it releases energy and warms the local environment.

The Water Cycle and the Landscape

The water cycle also influences the physical geography of the Earth. Glacial melt and erosion caused by water are two of the ways the water cycle helps create Earth's physical features.

As glaciers slowly expand across a landscape, they can carve away entire valleys, create mountain peaks, and leave behind rubble as big as boulders. Yosemite Valley, part of Yosemite National Park in the U.S. state of California, is a glacial valley. The famous Matterhorn, a peak on the Alps between Switzerland and Italy, was carved as glaciers collided and squeezed up the earth between them. Canada's "Big Rock" is one of the world's largest "glacial erratics," boulders left behind as a glacier advances or retreats.

Glacial melt can also create landforms. The Great Lakes, for example, are part of the landscape of the Midwest of the United States and Canada. The Great Lakes were created as an enormous ice sheet melted and retreated, leaving liquid pools.

The process of erosion and the movement of runoff also create varied landscapes across the Earth's surface. Erosion is the process by which earth is worn away by liquid water, wind, or ice.

Erosion can include the movement of runoff. The flow of water can help carve enormous canyons, for example. These canyons can be carved by rivers on high plateaus (such as the Grand Canyon, on the Colorado Plateau in the U.S. state of Arizona). They can also be carved by currents deep in the ocean (such as the Monterey Canyon, in the Pacific Ocean off the coast of the U.S. state of California).

Carbon Cycle

Carbon is a very important element, as it makes up organic matter, which is a part of all life. Carbon follows a certain route on earth, called the carbon cycle. Through following the carbon cycle we can also study energy flows on earth, because most of the chemical energy needed for life is stored in organic compounds as bonds between carbon atoms and other atoms. The carbon cycle naturally consists of two parts, the terrestrial and the aquatic carbon cycle. The aquatic carbon cycle is concerned with the movements of carbon through marine ecosystems and the terrestrial carbon cycle is concerned with the movement of carbon through terrestrial ecosystems.

The carbon cycle is based on carbon dioxide (CO_2), which can be found in air in the gaseous form, and in water in dissolved form. Terrestrial plants use atmospheric carbon dioxide from the atmosphere, to generate oxygen that sustains animal life. Aquatic plants also generate oxygen, but they use carbon dioxide from water. The process of oxygen generation is called photosynthesis. During photosynthesis, plants and other producers transfer carbon dioxide and water into complex carbohydrates, such as glucose, under the influence of sunlight. Only plants and some bacteria have the ability to conduct this process, because they possess chlorophyll; a pigment molecule in leaves that they can capture solar energy with.

The overall reaction of photosynthesis is: carbon dioxide + water + solar energy -> glucose + oxygen 6 CO_2 + 6 H_2O + solar energy -> $C_6H_{12}O_6$ + 6 O_2.

The oxygen that is produced during photosynthesis will sustain non-producing life forms, such as animals, and most micro organisms. Animals are called consumers, because they use the oxygen that is produced by plants. Carbon dioxide is released back into the atmosphere during respiration of consumers, which breaks down glucose and other complex organic compounds and converts the carbon back to carbon dioxide for reuse by producers.

Carbon that is used by producers, consumers and decomposers cycles fairly rapidly through air, water and biota. But carbon can also be stored as biomass in the roots of trees and other organic matter for many decades. This carbon is released back into the atmosphere by decomposition, as was noted before. Not all organic matter is immediately decomposed. Under certain conditions dead plant matter accumulates faster than it is decomposed within an ecosystem. The remains are locked away in underground deposits. When layers of sediment compress this matter fossil fuels will be formed, after many centuries. Long-term geological processes may expose the carbon in these fuels to air after a long period of time, but usually the carbon within the fossil fuels is released during humane combustion processes. The combustion of fossil fuels has supplied us with energy for as long as we can remember. But the human population of the world has been expanding and so has our demand for energy. That is why fossil fuels are burned very extensively. This is not without consequences, because we are burning fossil fuels much faster than they develop. Because of our actions fossil fuels have become non-renewable recourses.

Although the combustion of fossil fuels mainly adds carbon dioxide to air, some of it is also released during natural processes, such as volcanic eruptions.

In the aquatic ecosystem carbon dioxide can be stored in rocks and sediments. It will take a long time before this carbon dioxide will be released, through weathering of rocks or geologic processes that bring sediment to the surface of water. Carbon dioxide that is stored in water will be present as either carbonate or bicarbonate ions. These ions are an important part of natural buffers that prevent the water from becoming too acidic or too basic. When the sun warms up the water carbonate and bicarbonate ions will be returned to the atmosphere as carbon dioxide.

Main Components

Carbon pools in the major reservoirs on earth	
Pool	Quantity (gigatons)
Atmosphere	720
Ocean (total)	38,400
Total inorganic	37,400
Total organic	1,000
Surface layer	670
Deep layer	36,730
Lithosphere	
Sedimentary carbonates	> 60,000,000
Kerogens	15,000,000
Terrestrial biosphere (total)	2,000
Living biomass	600 - 1,000
Dead biomass	1,200

Aquatic biosphere	1 - 2
Fossil fuels (total)	4,130
Coal	3,510
Oil	230
Gas	140
Other (peat)	250

The global carbon cycle is now usually divided into the following major reservoirs of carbon interconnected by pathways of exchange:

- The atmosphere.
- The terrestrial biosphere.
- The ocean, including dissolved inorganic carbon and living and non-living marine biota.
- The sediments, including fossil fuels, fresh water systems and non-living organic material.
- The Earth's interior (mantle and crust. These carbon stores interact with the other components through geological processes.

The carbon exchanges between reservoirs occur as the result of various chemical, physical, geological, and biological processes. The ocean contains the largest active pool of carbon near the surface of the Earth. The natural flows of carbon between the atmosphere, ocean, terrestrial ecosystems, and sediments is fairly balanced, so that carbon levels would be roughly stable without human influence.

Atmosphere

Epiphytes on electric wires. This kind of plant takes both CO_2 and water from the atmosphere for living and growing.

Carbon in the Earth's atmosphere exists in two main forms: carbon dioxide and methane. Both of these gases absorb and retain heat in the atmosphere and are partially responsible for the greenhouse effect. Methane produces a larger greenhouse effect per volume as compared to carbon dioxide, but it exists in much lower concentrations and is more short-lived than carbon dioxide, making carbon dioxide the more important greenhouse gas of the two.

Carbon dioxide is removed from the atmosphere primarily through photosynthesis and enters the terrestrial and oceanic biospheres. Carbon dioxide also dissolves directly from the atmosphere into bodies of water (ocean, lakes, etc.), as well as dissolving in precipitation as raindrops fall through the atmosphere. When dissolved in water, carbon dioxide reacts with water molecules and forms carbonic acid, which contributes to ocean acidity. It can then be absorbed by rocks through weathering. It also can acidify other surfaces it touches or be washed into the ocean.

Human activities over the past two centuries have significantly increased the amount of carbon in the atmosphere, mainly in the form of carbon dioxide, both by modifying ecosystems' ability to extract carbon dioxide from the atmosphere and by emitting it directly, e.g., by burning fossil fuels and manufacturing concrete.

Terrestrial Biosphere

The terrestrial biosphere includes the organic carbon in all land-living organisms, both alive and dead, as well as carbon stored in soils. About 500 gigatons of carbon are stored above ground in plants and other living organisms, while soil holds approximately 1,500 gigatons of carbon. Most carbon in the terrestrial biosphere is organic carbon, while about a third of soil carbon is stored in inorganic forms, such as calcium carbonate. Organic carbon is a major component of all organisms living on earth. Autotrophs extract it from the air in the form of carbon dioxide, converting it into organic carbon, while heterotrophs receive carbon by consuming other organisms.

A portable soil respiration system measuring soil CO_2 flux.

Because carbon uptake in the terrestrial biosphere is dependent on biotic factors, it follows a diurnal and seasonal cycle. In CO_2 measurements, this feature is apparent in the Keeling curve. It is strongest in the northern hemisphere, because this hemisphere has more land mass than the southern hemisphere and thus more room for ecosystems to absorb and emit carbon.

Carbon leaves the terrestrial biosphere in several ways and on different time scales. The combustion or respiration of organic carbon releases it rapidly into the atmosphere. It can also be exported into the ocean through rivers or remain sequestered in soils in the form of inert carbon. Carbon stored in soil can remain there for up to thousands of years before being washed into rivers by erosion or released into the atmosphere through soil respiration. Between 1989 and 2008 soil respiration increased by about 0.1% per year. In 2008, the global total of CO_2 released by soil respiration was roughly 98 billion tonnes, about 10 times more carbon than humans are now putting into the atmosphere each year by burning fossil fuel (this does not represent a net transfer of carbon from soil to atmosphere, as the respiration is largely offset by inputs to soil carbon). There are a few plausible explanations for this trend, but the most likely explanation is that increasing temperatures have increased rates of decomposition of soil organic matter, which has increased the flow of CO_2. The length of carbon sequestering in soil is dependent on local climatic conditions and thus changes in the course of climate change.

Ocean

The ocean can be conceptually divided into a surface layer within which water makes frequent (daily to annual) contact with the atmosphere, and a deep layer below the typical mixed layer depth of a few hundred meters or less, within which the time between consecutive contacts may be centuries. The dissolved inorganic carbon (DIC) in the surface layer is exchanged rapidly with the atmosphere, maintaining equilibrium. Partly because its concentration of DIC is about 15% higher but mainly due to its larger volume, the deep ocean contains far more carbon--it›s the largest pool of actively cycled carbon in the world, containing 50 times more than the atmosphere--but the timescale to reach equilibrium with the atmosphere is hundreds of years: the exchange of carbon between the two layers, driven by thermohaline circulation, is slow.

Carbon enters the ocean mainly through the dissolution of atmospheric carbon dioxide, a small fraction of which is converted into carbonate. It can also enter the ocean through rivers as dissolved organic carbon. It is converted by organisms into organic carbon through photosynthesis and can either be exchanged throughout the food chain or precipitated into the ocean›s deeper, more carbon rich layers as dead soft tissue or in shells as calcium carbonate. It circulates in this layer for long periods of time before either being deposited as sediment or, eventually, returned to the surface waters through thermohaline circulation.

Oceanic absorption of CO_2 is one of the most important forms of carbon sequestering limiting the human-caused rise of carbon dioxide in the atmosphere. However, this process is limited by a number of factors. CO_2 absorption makes water more acidic, which affects ocean biosystems. The projected rate of increasing oceanic acidity could slow the biological precipitation of calcium carbonates, thus decreasing the ocean's capacity to absorb carbon dioxide.

Earth's Interior

The geologic component of the carbon cycle operates slowly in comparison to the other parts of the global carbon cycle. It is one of the most important determinants of the amount of carbon in the atmosphere, and thus of global temperatures.

Most of the earth's carbon is stored inertly in the earth's lithosphere. Much of the carbon stored in the earth›s mantle was stored there when the earth formed. Some of it was deposited in the form of organic carbon from the biosphere. Of the carbon stored in the geosphere, about 80% is limestone and its derivatives, which form from the sedimentation of calcium carbonate stored in the shells of marine organisms. The remaining 20% is stored as kerogens formed through the sedimentation and burial of terrestrial organisms under high heat and pressure. Organic carbon stored in the geosphere can remain there for millions of years.

Carbon can leave the geosphere in several ways. Carbon dioxide is released during the metamorphosis of carbonate rocks when they are subducted into the earth›s mantle. This carbon dioxide can be released into the atmosphere and ocean through volcanoes and hotspots. It can also be removed by humans through the direct extraction of kerogens in the form of fossil fuels. After extraction, fossil fuels are burned to release energy, thus emitting the carbon they store into the atmosphere.

Human Influence

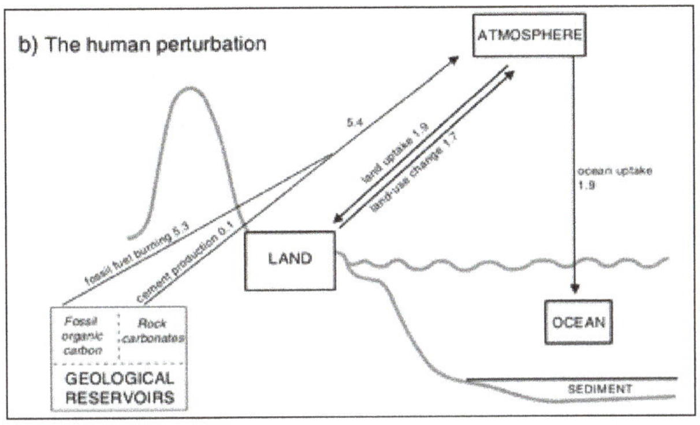

Human activity since the industrial era has changed the balance in the natural carbon cycle. Units are in gigatons.

CO_2 in Earth's atmosphere if half of global-warming emissions are notabsorbed.

Since the industrial revolution, human activity has modified the carbon cycle by changing its components' functions and directly adding carbon to the atmosphere.

The largest human impact on the carbon cycle is through direct emissions from burning fossil fuels, which transfers carbon from the geosphere into the atmosphere. The rest of this increase is caused mostly by changes in land-use, particularly deforestation.

Another direct human impact on the carbon cycle is the chemical process of calcination of limestone for clinker production, which releases CO_2. Clinker is an industrial precursor of cement.

Humans also influence the carbon cycle indirectly by changing the terrestrial and oceanic biosphere. Over the past several centuries, direct and indirect human-caused land use and land cover change (LUCC) has led to the loss of biodiversity, which lowers ecosystems' resilience to environmental stresses and decreases their ability to remove carbon from the atmosphere. More directly, it often leads to the release of carbon from terrestrial ecosystems into the atmosphere. Deforestation for agricultural purposes removes forests, which hold large amounts of carbon, and replaces them, generally with agricultural or urban areas. Both of these replacement land cover types store comparatively small amounts of carbon, so that the net product of the process is that more carbon stays in the atmosphere.

Other human-caused changes to the environment change ecosystems' productivity and their ability to remove carbon from the atmosphere. Air pollution, for example, damages plants and soils, while many agricultural and land use practices lead to higher erosion rates, washing carbon out of soils and decreasing plant productivity.

Humans also affect the oceanic carbon cycle. Current trends in climate change lead to higher ocean temperatures, thus modifying ecosystems. Also, acid rain and polluted runoff from agriculture and industry change the ocean's chemical composition. Such changes can have dramatic effects on highly sensitive ecosystems such as coral reefs, thus limiting the ocean's ability to absorb carbon from the atmosphere on a regional scale and reducing oceanic biodiversity globally.

Arctic methane emissions indirectly caused by anthropogenic global warming also affect the carbon cycle, and contribute to further warming in what is known as climate change feedback.

On 12 November 2015, NASA scientists reported that human-made carbon dioxide (CO_2) continues to increase, reaching levels not seen in hundreds of thousands of years: currently, the rate carbon dioxide release by the burning of fossil fuels is about double the net uptake by vegetation and the ocean.

The carbon cycle is most easily studied as two interconnected subcycles:

- One dealing with rapid carbon exchange among living organisms.
- One dealing with long-term cycling of carbon through geologic processes.

Although we will look at them separately, it's important to realize these cycles are linked. For instance, the same pools of atmospheric and oceanic CO_2 that are utilized by organisms are also fed and depleted by geological processes.

As a brief overview, carbon exists in the air largely as carbon dioxide—CO_2—gas, which dissolves in water and reacts with water molecules to produce bicarbonate—HCO_3^-. Photosynthesis by land plants, bacteria, and algae converts carbon dioxide or bicarbonate into organic molecules. Organic molecules made by photosynthesizers are passed through food chains, and cellular respiration converts the organic carbon back into carbon dioxide gas.

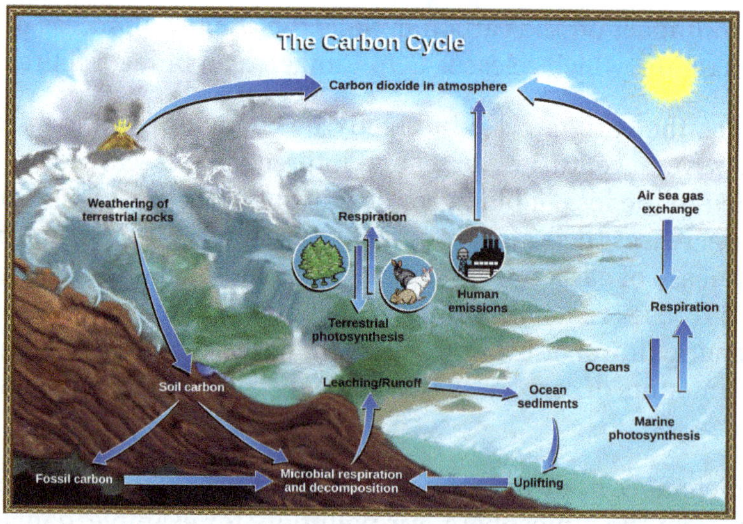

Longterm storage of organic carbon occurs when matter from living organisms is buried deep underground or sinks to the bottom of the ocean and forms sedimentary rock. Volcanic activity and, more recently, human burning of fossil fuels bring this stored carbon back into the carbon cycle. Although the formation of fossil fuels happens on a slow, geologic timescale, human release of the carbon they contain—as CO_2—is on a very fast timescale.

The Biological Carbon Cycle

Carbon enters all food webs, both terrestrial and aquatic, through autotrophs, or self-feeders. Almost all of these autotrophs are photosynthesizers, such as plants or algae.

Autotrophs capture carbon dioxide from the air or bicarbonate ions from the water and use them to make organic compounds such as glucose. Heterotrophs, or other-feeders, such as humans, consume the organic molecules, and the organic carbon is passed through food chains and webs.

How does carbon cycle back to the atmosphere or ocean? To release the energy stored in carbon-containing molecules, such as sugars, autotrophs and heterotrophs break these molecules down in a process called cellular respiration. In this process, the carbons of the molecule are released as carbon dioxide. Decomposers also release organic compounds and carbon dioxide when they break down dead organisms and waste products.

Carbon can cycle quickly through this biological pathway, especially in aquatic ecosystems. Overall, an estimated 1,000 to 100,000 million metric tons of carbon move through the biological pathway each year. For context, a metric ton is about the weight of an elephant or a small car!

The Geological Carbon Cycle

The geological pathway of the carbon cycle takes much longer than the biological pathway described above. In fact, it usually takes millions of years for carbon to cycle through the geological pathway. Carbon may be stored for long periods of time in the atmosphere, bodies of liquid water—mostly oceans— ocean sediment, soil, rocks, fossil fuels, and Earth's interior.

The level of carbon dioxide in the atmosphere is influenced by the reservoir of carbon in the oceans and vice versa. Carbon dioxide from the atmosphere dissolves in water and reacts with water molecules in the following reactions:

$$CO_2 + H_2O \rightleftharpoons H_2CO_3 \rightleftharpoons HCO_3^- + H^+ \rightleftharpoons CO_3^{2-} + 2H^+$$

The carbonate— CO_3^{2-} —released in this process combines with Ca^{2+} ions to make calcium carbonate— $CaCO_3$ —a key component of the shells of marine organisms. When the organisms die, their remains may sink and eventually become part of the sediment on the ocean floor. Over geologic time, the sediment turns into limestone, which is the largest carbon reservoir on Earth.

On land, carbon is stored in soil as organic carbon from the decomposition of living organisms or as inorganic carbon from weathering of terrestrial rock and minerals. Deeper under the ground are fossil fuels such as oil, coal, and natural gas, which are the remains of plants decomposed under anaerobic—oxygen-free—conditions. Fossil fuels take millions of years to form. When humans burn them, carbon is released into the atmosphere as carbon dioxide.

Another way for carbon to enter the atmosphere is by the eruption of volcanoes. Carbon-containing sediments in the ocean floor are taken deep within the Earth in a process called subduction, in which one tectonic plate moves under another. This process forms carbon dioxide, which can be released into the atmosphere by volcanic eruptions or hydrothermal vents.

Oxygen Cycle

The oxygen cycle elaborates how oxygen circulates in various forms through nature. Oxygen occurs freely in the air, trapped in the earth crust as chemical compounds, or dissolved in water. Oxygen in the atmosphere is about 21%, and it is the second most abundant gas after nitrogen. It is mostly utilized by living organisms, especially man and animals in respiration. Oxygen is also the most common element of human body.

Oxygen is also used during combustion, decomposition, and oxidation. The circulation of oxygen is through three main flow systems including the (air) atmosphere, the biosphere, and the earth's crust. In the oxygen cycle, the main driving factor is photosynthesis which is the process whereby green plants and algae make their own food by use of solar energy, water, and carbon dioxide to gives off oxygen as a by-product.

Hence, for oxygen to remain in the atmosphere, it has to circulate through various forms of nature which is essentially termed as the oxygen cycle. The circulation depends on the various activities on Earth.

Reservoirs

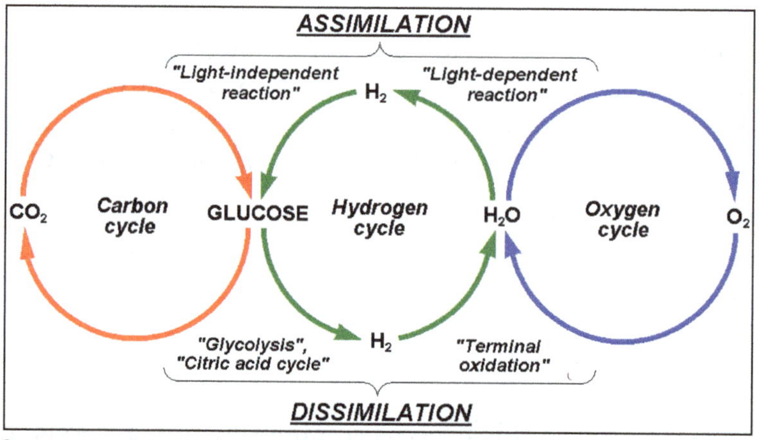

Interconnection between carbon, hydrogen and oxygen cycle in metabolism of photosynthesizing plants.

By far the largest reservoir of Earth's oxygen is within the silicate and oxide minerals of the crustand mantle (99.5% by weight). The Earth›s atmosphere, hydrosphere and biosphere together weigh less than 0.05% of the Earth›s total mass. Oxygen is one of the most abundant elements on Earth and represents a large portion of each main reservoir:

- Atmosphere is 21% oxygen by volume present mainly as free oxygen molecules (O_2) with other oxygen-containing molecules including ozone (O_3), carbon dioxide (CO_2), water vapor (H_2O), and sulfur and nitrogen oxides (SO_2, NO, N_2O, etc.).
- Biosphere is 22% oxygen by volume present mainly as a component of organic molecules ($C_xH_xN_xO_x$) and water molecules.
- Hydrosphere is 33% oxygen by volume present mainly as a component of water molecules with dissolved molecules including free oxygen and carbonic acids (H_xCO_3).

- Lithosphere is 94% oxygen by volume present mainly as silica minerals (SiO_2) and other oxide minerals.

Movement of oxygen between the reservoirs is facilitated in large part by the presence of atmospheric free oxygen. The main source of atmospheric free oxygen is photosynthesis, which produces sugars and free oxygen from carbon dioxide and water:

$$6\,CO_2 + 6H_2O + energy \rightarrow C_6H_{12}O_6 + 6\,O_2$$

Photosynthesizing organisms include the plant life of the land areas as well as the phytoplankton of the oceans. The tiny marine cyanobacterium Prochlorococcus was discovered in 1986 and accounts for more than half of the photosynthesis of the open ocean.

An additional source of atmospheric free oxygen comes from photolysis, whereby high-energy ultraviolet radiation breaks down atmospheric water and nitrous oxide into component atoms. The free H and N atoms escape into space, leaving O_2 in the atmosphere:

$$2H_2O + energy \rightarrow 4H + O_2$$

$$2N_2O + energy \rightarrow 4N + O_2$$

The main way free oxygen is lost from the atmosphere is via respiration and decay, mechanisms in which animal life and bacteria consume oxygen and release carbon dioxide.

The lithosphere also consumes atmospheric free oxygen by chemical weathering and surface reactions. An example of surface weathering chemistry is formation of iron oxides (rust):

$$4\,FeO + O_2 \rightarrow 2\,Fe_2O_3$$

Oxygen is also cycled between the biosphere and lithosphere. Marine organisms in the biosphere create calcium carbonate shell material ($CaCO_3$) that is rich in oxygen. When the organism dies, its shell is deposited on the shallow sea floor and buried over time to create the limestone sedimentary rock of the lithosphere. Weathering processes initiated by organisms can also free oxygen from the lithosphere. Plants and animals extract nutrient minerals from rocks and release oxygen in the process.

Capacities and Fluxes

The following tables offer estimates of oxygen cycle reservoir capacities and fluxes. These numbers are based primarily on estimates from (Walker, J. C. G.):

Table: Major reservoirs involved in the oxygen cycle

Reservoir	Capacity (kg O_2)	Flux in/out (kg O_2 per year)	Residence time (years)
Atmosphere	1.4×10^{18}	3×10^{14}	4500
Biosphere	1.6×10^{16}	3×10^{14}	50
Lithosphere	2.9×10^{20}	6×10^{11}	500000000

Table : Annual gain and loss of atmospheric oxygen (Units of 10^{10} kg O_2 per year)

Photosynthesis (land) Photosynthesis (ocean) Photolysis of N_2O Photolysis of H_2O	16,500 13,500 1.3 0.03
Total gains	~ 30,000
Losses - respiration and decay	
Aerobic respiration Microbial oxidation Combustion of fossil fuel (anthropogenic) Photochemical oxidation Fixation of N_2 by lightning Fixation of N_2 by industry (anthropogenic) Oxidation of volcanic gases	23,000 5,100 1,200 600 12 10 5
Losses - weathering	
Chemical weathering Surface reaction of O_3	50 12
Total losses	~ 30,000

Ozone

The presence of atmospheric oxygen has led to the formation of ozone (O_3) and the ozone layer within the stratosphere:

$$O_2 + uv\,light \rightarrow 2O \qquad (\lambda \lesssim 200\,nm)$$

$$O + O_2 \rightarrow O_3$$

The ozone layer is extremely important to modern life as it absorbs harmful ultraviolet radiation:

$$O_3 + uv\,light \rightarrow O_2 + O \qquad (\lambda \lesssim 300\,nm)$$

Process of Oxygen Cycle

The Atmosphere (air)

The atmosphere carries a small quantity of all oxygen, only about 0.35% of the entire earth's oxygen. In the atmosphere, oxygen is released by the process known as photolysis. Photolysis happens when the ultraviolet radiation of sunlight breaks apart oxygen-containing molecules such as nitrous oxide and atmospheric water to release free oxygen. The surplus oxygen recombines with other oxygen molecules to form ozone while the rest is freed into the atmosphere. Ozone is the layer that helps to shield the Earth from the dangerous ultra violet rays.

Biosphere

The biosphere carries the smallest quantity of all earth's oxygen, about 0.01%. In the biosphere, the major oxygen cycles are photosynthesis and respiration. In these two processes of the oxygen cycle, it is interconnected with the carbon cycle and the water cycle. During photosynthesis, plants and planktons use sunlight energy, water, and carbon dioxide to make food (carbohydrates) and release oxygen as a by-product. As such, plants and planktons are the main producers of oxygen in the ecosystem. They take in carbon dioxide and give out oxygen. Plants are estimated to replace about 99% of all the oxygen used.

On the other hand, respiration happens when humans and animals breathe in oxygen which is used during metabolism to break down carbohydrates and exhale carbon dioxide as a by-product. Such free carbon dioxide is then released into the environment and is used by plants and planktons during photosynthesis to give out molecular atmospheric oxygen, thus completing the oxygen cycle. Therefore, suffice is to say that oxygen enters organisms in the biosphere through respiration and is expelled through photosynthesis in a process that is interconnected with the carbon cycle-plus the water cycle.

However, the continued release of carbon dioxide into the atmosphere by burning fossil fuels and automobile pollution affects the oxygen cycle.

The Lithosphere (Earth's Crust)

The lithosphere carries the largest quantity of all earth's oxygen, about 99.5%, because it is a constituent of the earth's lands, soils, organic matter, biomass, water, and rocks. Mostly, these constituents of the earth fix oxygen in mineral chemicals compounds such as oxides and silicates. The process is natural and happens automatically as the pure mineral elements absorb or react with the free oxygen. It happens similar to the manner in which iron picks up oxygen from the air, resulting in the formation of rust (iron oxides).

As such, during chemical reactions and some weathering processes, a portion of the trapped oxygen in the minerals is released into the atmosphere. Also, as animals and plants draw nutrient minerals from rocks, organic matter, or biomass, some of the trapped oxygen is freed in the process. Dissolved oxygen is also present in water system which is essential for the survival of aquatic life forms. As a result, these processes combined gives rise to oxygen cycle in the biosphere and lithosphere.

Processes that Use Oxygen

1. Respiration: When we breather, we use oxygen and release carbon dioxide. Similarly, animals and plants also use oxygen when they breathe.
2. Combustion: When you burn something let's say a paper, you need three things for combustion to take place i.e. oxygen, fuel and heat. So, when you burn a paper, it uses oxygen and releases carbon dioxide and may be some other gases.
3. Decomposition: Decomposition occurs when plants and animals die. When this happens, they decompose and such process uses oxygen and releases carbon dioxide.
4. Rusting: When things rust, they use oxygen. This is also called as oxidation.

Processes that Produce Oxygen

1. Plants: Plants produce oxygen during the process of photosynthesis. During the process of photosynthesis, plants use carbon dioxide, sunlight, and water to create energy.
2. Sunlight: Some oxygen also gets produced when sunlight reacts with water vapor in the atmosphere.

Nitrogen Cycle

The nitrogen cycle refers to the cycle of nitrogen atoms through the living and non-living systems of Earth. Because nitrogen is a necessary ingredient for life as we know it, the nitrogen cycle is vital to sustaining life on Earth.

Nitrogen was originally formed in the hearts of stars through the process of nuclear fusion. When ancient stars exploded, they flung nitrogen-containing gases across the Universe. When the Earth was formed, nitrogen gas was a main ingredient in its atmosphere.

Today, the Earth's atmosphere is about 78% nitrogen, about 21% oxygen, and about 1% other gases. This is an ideal balance because too much oxygen can actually be toxic to cells, as well as being highly flammable. Nitrogen, on the other hand, is inert and harmless in its gaseous form.

Here we will discuss how nitrogen plays a vital role in the chemistry of life – and how it gets from the atmosphere, into living things, and back again.

Function of Nitrogen Cycle

Nitrogen is an essential ingredient for life as we know it. Its unique chemical bonding properties allow it to create structures such as DNA and RNA nucleotides, and the amino acids from which proteins are built. Without nitrogen, these molecules would not be able to exist.

It's thought that the first nucleotides and amino acids formed naturally under the volatile conditions of early Earth, where energy sources like lightning strikes could cause nitrogen and other atoms to react and form complex structures.

This process might have naturally produced self-replicating organic chemicals – but in order to reproduce and evolve, life needed to figure out how to make these nitrogen compounds on demand.

Today, "nitrogen fixers" are organisms that can turn nitrogen gas from the atmosphere into nitrogen compounds that other organisms can use to produce nucleic acids, amino acids, and more. These nitrogen fixers are such a vital part of the ecosystem that agriculture cannot occur without them.

Ancient peoples learned that if they did not alternate growing nitrogen-consuming crops with nitrogen-fixing crops, their farms would become fallow and unable to support growth. Today, most artificial fertilizers contain life-giving nitrogen compounds as their main ingredient to make soil more fertile.

Processes

Nitrogen is present in the environment in a wide variety of chemical forms including organic nitrogen, ammonium (NH^+_4), nitrite(NO^-_2), nitrate (NO^-_3), nitrous oxide (N_2O), nitric oxide (NO) or inorganic nitrogen gas (N_2). Organic nitrogen may be in the form of a living organism, humusor in the intermediate products of organic matter decomposition. The processes of the nitrogen cycle transform nitrogen from one form to another. Many of those processes are carried out by microbes, either in their effort to harvest energy or to accumulate nitrogen in a form needed for their growth. For example, the nitrogenous wastes in animal urine are broken down by nitrifying bacteria in the soil to be used as new. The diagram {besides} alongside shows how these processes fit together to form the nitrogen cycle.

Nitrogen Fixation

The conversion of nitrogen gas (N_2) into nitrates and nitrites through atmospheric, industrial and biological processes is called nitrogen fixation. Atmospheric nitrogen must be processed, or "fixed", into a usable form to be taken up by plants. Between 5×10^{12} and 10×10^{12} g per year are fixed by lightning strikes, but most fixation is done by free-living or symbiotic bacteria known as diazotrophs. These bacteria have the nitrogenase enzyme that combines gaseous nitrogen with hydrogen to produce ammonia, which is converted by the bacteria into other organic compounds. Most biological nitrogen fixation occurs by the activity of Mo-nitrogenase, found in a wide variety of bacteria and some Archaea. Mo-nitrogenase is a complex two-component enzyme that has multiple metal-containing prosthetic groups. An example of free-living bacteria is Azotobacter. Symbiotic nitrogen-fixing bacteria such as Rhizobium usually live in the root nodules of legumes (such as peas, alfalfa, and locust trees). Here they form a mutualistic relationship with the plant, producing ammonia in exchange for carbohydrates. Because of this relationship, legumes will often increase the nitrogen content of nitrogen-poor soils. A few non-legumes can also form such symbioses. Today, about 30% of the total fixed nitrogen is produced industrially using the Haber-Bosch process, which uses high temperatures and pressures to convert nitrogen gas and a hydrogen source (natural gas or petroleum) into ammonia.

Assimilation

Plants can absorb nitrate or ammonium from the soil via their root hairs. If nitrate is absorbed, it is first reduced to nitrite ions and then ammonium ions for incorporation into amino acids, nucleic acids, and chlorophyll. In plants that have a symbiotic relationship with rhizobia, some nitrogen is assimilated in the form of ammonium ions directly from the nodules. It is now known that there is a more complex cycling of amino acids between Rhizobia bacteroids and plants. The plant provides amino acids to the bacteroids so ammonia assimilation is not required and the bacteroids pass amino acids (with the newly fixed nitrogen) back to the plant, thus forming an interdependent relationship. While many animals, fungi, and other heterotrophicorganisms obtain nitrogen by ingestion of amino acids, nucleotides, and other small organic molecules, other heterotrophs (including many bacteria) are able to utilize inorganic compounds, such as ammonium as sole N sources. Utilization of various N sources is carefully regulated in all organisms.

Ammonification

When a plant or animal dies or an animal expels waste, the initial form of nitrogen is organic. Bacteria or fungi convert the organic nitrogen within the remains back into ammonium ($NH^+{}_4$), a process called ammonification or mineralization. Enzymes involved are:

- GS: Gln Synthetase (Cytosolic & Plastic).
- GOGAT: Glu 2-oxoglutarate aminotransferase (Ferredoxin & NADH-dependent).
- GDH: Glu Dehydrogenase:

- o Minor Role in ammonium assimilation.
- o Important in amino acid catabolism.

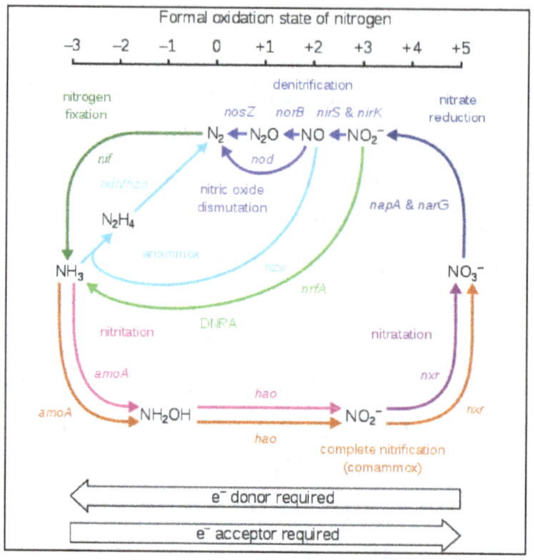

A schematic representation of the microbial nitrogen cycle. ANAMMOX is anaerobic ammonium oxidation, DNRA is dissimilatory nitrate reduction to ammonium, and COMMAMOXis complete ammonium oxidation.

The conversion of ammonium to nitrate is performed primarily by soil-living bacteria and other nitrifying bacteria. In the primary stage of nitrification, the oxidation of ammonium (NH^{+}_{4}) is performed by bacteria such as the Nitrosomonas species, which converts ammonia to nitrites (NO^{-}_{2}). Other bacterial species such as Nitrobacter, are responsible for the oxidation of the nitrites (NO^{-}_{2}) into nitrates (NO^{-}_{3}). It is important for the ammonia (NH_3) to be converted to nitrates or nitrites because ammonia gas is toxic to plants.

Due to their very high solubility and because soils are highly unable to retain anions, nitrates can enter groundwater. Elevated nitrate in groundwater is a concern for drinking water use because nitrate can interfere with blood-oxygen levels in infants and cause methemoglobinemia or blue-baby syndrome. Where groundwater recharges stream flow, nitrate-enriched groundwater can contribute to eutrophication, a process that leads to high algal population and growth, especially blue-green algal populations. While not directly toxic to fish life, like ammonia, nitrate can have indirect effects on fish if it contributes to this eutrophication. Nitrogen has contributed to severe eutrophication problems in some water bodies. Since 2006, the application of nitrogen fertilizer has been increasingly controlled in Britain and the United States. This is occurring along the same lines as control of phosphorus fertilizer, restriction of which is normally considered essential to the recovery of eutrophied waterbodies.

Denitrification

Denitrification is the reduction of nitrates back into nitrogen gas (N_2), completing the nitrogen cycle. This process is performed by bacterial species such as Pseudomonas and Clostridium in anaerobic conditions. They use the nitrate as an electron acceptor in the place of oxygen during respiration. These facultatively anaerobic bacteria can also live in aerobic conditions. Denitrification happens in anaerobic conditions e.g. waterlogged soils. The denitrifying bacteria use nitrates in the soil to carry out respiration and consequently produce nitrogen gas, which is inert and unavailable to plants.

Dissimilatory Nitrate Reduction to Ammonium

Dissimilatory nitrate reduction to ammonium (DNRA), or nitrate/nitrite ammonification, is an anaerobic respiration process. Microbes which undertake DNRA oxidise organic matter and use nitrate as an electron acceptor, reducing it to nitrite, then ammonium ($NO_3^- \rightarrow NO_2^- \rightarrow NH_4^+$). Both denitrifying and nitrate ammonification bacteria will be competing for nitrate in the environment, although DNRA acts to conserve bioavailable nitrogen as soluble ammonium rather than producing dinitrogen gas.

Anaerobic Ammonia Oxidation

In this biological process, nitrite and ammonia are converted directly into molecular nitrogen (N_2) gas. This process makes up a major proportion of nitrogen conversion in the oceans. The balanced formula for this "anammox" chemical reaction is: $NH^+_4 + NO^-_2 \rightarrow N_2 + 2H_2O$ ($\Delta G° = -357$ kJ mol^{-1}).

Other Processes

Though nitrogen fixation is the primary source of plant-available nitrogen in most ecosystems, in areas with nitrogen-rich bedrock, the breakdown of this rock also serves as a nitrogen source.

Marine Nitrogen Cycle

The nitrogen cycle is an important process in the ocean as well. While the overall cycle is similar, there are different players and modes of transfer for nitrogen in the ocean. Nitrogen enters the water through the precipitation, runoff, or as N_2 from the atmosphere. Nitrogen cannot be utilized by phytoplankton as N_2 so it must undergo nitrogen fixation which is performed predominately by cyanobacteria. Without supplies of fixed nitrogen entering the marine cycle, the fixed nitrogen would be used up in about 2000 years.Phytoplankton need nitrogen in biologically available forms for the initial synthesis of organic matter. Ammonia and urea are released into the water by excretion from plankton. Nitrogen sources are removed from the euphotic zone by the downward movement of the organic matter. This can occur from sinking of phytoplankton, vertical mixing, or sinking of waste of vertical migrators. The sinking results in ammonia being introduced at lower depths below the euphotic zone. Bacteria are able to convert ammonia to nitrite and nitrate but they are inhibited by light so this must occur below the euphotic zone. Ammonification or Mineralization is performed by bacteria to convert organic nitrogen to ammonia. Nitrification can then occur to convert the ammonium to nitrite and nitrate. Nitrate can be returned to the euphotic zone by vertical mixing and upwelling where it can be taken up by phytoplankton to continue the cycle. N_2 can be returned to the atmosphere through denitrification.

Ammonium is thought to be the preferred source of fixed nitrogen for phytoplankton because its assimilation does not involve a redox reaction and therefore requires little energy. Nitrate requires a redox reaction for assimilation but is more abundant so most phytoplankton have adapted to have the enzymes necessary to undertake this reduction (nitrate reductase). There are a few notable and well-known exceptions that include Prochlorococcus and some Synechococcus. These species can only take up nitrogen as ammonium.

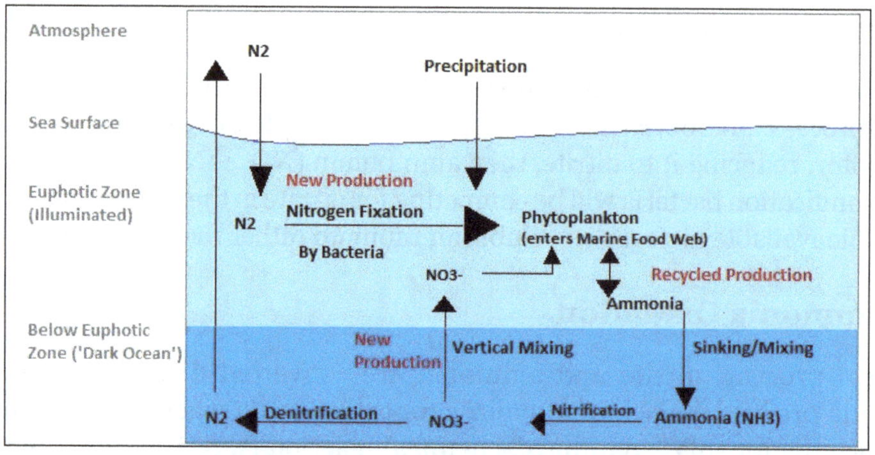

A schematic representing the Marine Nitrogen Cycle.

The nutrients in the ocean are not uniformly distributed. Areas of upwelling provide supplies of nitrogen from below the euphotic zone. Coastal zones provide nitrogen from runoff and upwelling occurs readily along the coast. However, the rate at which nitrogen can be taken up by phytoplankton is decreased in oligotrophic waters year-round and temperate water in the summer resulting in lower primary production.The distribution of the different forms of nitrogen varies throughout the oceans as well.

Nitrate is depleted in near-surface water except in upwelling regions. Coastal upwelling regions usually have high nitrate and chlorophyll levels as a result of the increased production. However, there are regions of high surface nitrate but low chlorophyll that are referred to as HNLC (high nitrogen, low chlorophyll) regions. The best explanation for HNLC regions relates to iron scarcity in the ocean, which may play an important part in ocean dynamics and nutrient cycles. The input of iron varies by region and is delivered to the ocean by dust (from dust storms) and leached out of rocks. Iron is under consideration as the true limiting element to ecosystem productivity in the ocean.

Ammonium and nitrite show a maximum concentration at 50–80 m (lower end of the euphotic zone) with decreasing concentration below that depth. This distribution can be accounted for by the fact that nitrite and ammonium are intermediate species. They are both rapidly produced and consumed through the water column. The amount of ammonium in the ocean is about 3 orders of magnitude less than nitrate. Between ammonium, nitrite, and nitrate, nitrite has the fastest turnover rate. It can be produced during nitrate assimilation, nitrification, and denitrification; however, it is immediately consumed again.

New vs. Regenerated Nitrogen

Nitrogen entering the euphotic zone is referred to as new nitrogen because it is newly arrived from outside the productive layer. The new nitrogen can come from below the euphotic zone or from outside sources. Outside sources are upwelling from deep water and nitrogen fixation. If the organic matter is eaten, respired, delivered to the water as ammonia, and re-incorporated into organic matter by phytoplankton it is considered recycled/regenerated production. New production is an important component of the marine environment. One reason is that only continual input of new nitrogen can determine the total capacity of the ocean to produce a sustainable fish harvest. Harvesting fish from regenerated nitrogen areas will lead to a decrease in nitrogen and therefore a decrease in primary production. This will have a negative effect on the system. However, if fish are harvested from areas of new nitrogen the nitrogen will be replenished.

Human Influences on the Nitrogen Cycle

As a result of extensive cultivation of legumes (particularly soy, alfalfa, and clover), growing use of the Haber–Bosch process in the creation of chemical fertilizers, and pollution emitted by vehicles and industrial plants, human beings have more than doubled the annual transfer of nitrogen into biologically available forms. In addition, humans have significantly contributed to the transfer of nitrogen trace gases from Earth to the atmosphere and from the land to aquatic systems. Human alterations to the global nitrogen cycle are most intense in developed countries and in Asia, where vehicle emissions and industrial agriculture are highest.

Nitrous oxide (N_2O) has risen in the atmosphere as a result of agricultural fertilization, biomass burning, cattle and feedlots, and industrial sources. N_2O has deleterious effects in the stratosphere, where it breaks down and acts as a catalyst in the destruction of atmospheric ozone. Nitrous oxide is also a greenhouse gas and is currently the third largest contributor to global warming, after carbon dioxide and methane. While not as abundant in the atmosphere as carbon dioxide, it is, for an equivalent mass, nearly 300 times more potent in its ability to warm the planet.

Ammonia (NH_3) in the atmosphere has tripled as the result of human activities. It is a reactant in the atmosphere, where it acts as an aerosol, decreasing air quality and clinging to water droplets, eventually resulting in nitric acid (HNO_3) that produces acid rain. Atmospheric ammonia and nitric acid also damage respiratory systems.

The very high temperature of lightning naturally produces small amounts of NO_x, NH_3, and HNO_3, but high-temperature combustion has contributed to a 6- or 7-fold increase in the flux of NO_x to the atmosphere. Its production is a function of combustion temperature - the higher the temperature, the more NO_x is produced. Fossil fuel combustion is a primary contributor, but so are biofuels and even the burning of hydrogen. However, the rate that hydrogen is directly injected into the combustion chambers of internal combustion engines can be controlled to prevent the higher combustion temperatures that produce NO_x.

Ammonia and nitrous oxides actively alter atmospheric chemistry. They are precursors of tropospheric (lower atmosphere) ozone production, which contributes to smog and acid rain, damages plants and increases nitrogen inputs to ecosystems. Ecosystem processes can increase with nitrogen fertilization, but anthropogenic input can also result in nitrogen saturation, which weakens productivity and can damage the health of plants, animals, fish, and humans.

Decreases in biodiversity can also result if higher nitrogen availability increases nitrogen-demanding grasses, causing a degradation of nitrogen-poor, species-diverse heathlands.

Environmental Impacts

Increasing levels of nitrogen deposition are shown to have a number of negative effects on both terrestrial and aquatic ecosystems. Nitrogen gases and aerosols can be directly toxic to certain plant species, affecting the aboveground physiology and growth of plants near large point sources of nitrogen pollution. Changes to plant species interactions may also occur, as accumulation of nitrogen compounds increase its availability in a given ecosystem, eventually changing the species composition, plant diversity, and nitrogen cycling. Ammonia and ammonium - two reduced forms of nitrogen - can be detrimental over time due to an increased toxicity toward sensitive species of plants, particularly those that are accustomed to using nitrate as their source of nitrogen, causing poor development of their roots and shoots. Increased nitrogen deposition also leads to soil acidification, which increases base cation leaching in the soil and amounts of aluminium and other potentially toxic metals, along with decreasing the amount of nitrification occurring and increasing plant-derived litter. Due to the ongoing changes caused by high nitrogen deposition, an environment›s susceptibility to ecological stress and disturbance - such as pests and pathogens - may increase, thus making it less resilient to situations that otherwise would have little impact to its long-term vitality.

Additional risks posed by increased availability of inorganic nitrogen in aquatic ecosystems include water acidification; eutrophication of fresh and saltwater systems; and toxicity issues for animals, including humans. Eutrophication often leads to lower dissolved oxygen levels in the water column, including hypoxic and anoxic conditions, which can cause death of aquatic fauna. Relatively sessile benthos, or bottom-dwelling creatures, are particularly vulnerable because of their lack of mobility, though large fish kills are not uncommon. Oceanic dead zones near the mouth of the Mississippi in the Gulf of Mexico are a well-known example of algal bloom-induced hypoxia. The New York Adirondack Lakes, Catskills, Hudson Highlands, Rensselaer Plateau and parts of Long Island display the impact of nitric acid rain deposition, resulting in the killing of fish and many other aquatic species.

Ammonia (NH_3) is highly toxic to fish and the level of ammonia discharged from wastewater treatment facilities must be closely monitored. To prevent fish deaths, nitrification via aeration prior to discharge is often desirable. Land application can be an attractive alternative to the aeration.

What is the importance of the nitrogen cycle?

- As we all know by now, the nitrogen cycle helps bring in the inert nitrogen from the air into the biochemical process in plants and then to animals.
- Plants need nitrogen to synthesise chlorophyll and so the nitrogen cycle is absolutely essential for them.
- During the process of ammonification, the bacteria help degrade decomposing animal and plant matter. This helps in naturally cleaning up the environment.
- Due to the nitrogen cycle, nitrates and nitrites are released into the soil which helps in enriching the soil with nutrients needed for cultivation.
- As plants use nitrogen for their biochemical processes, animals obtain the nitrogen and nitrogen compounds from plants. Nitrogen is needed as is an integral part of the cell composition. It is due to the nitrogen cycle that animals are also able to utilise the nitrogen present in the air.

Phosphorus Cycle

Phosphorus is an important element for all forms of life. As phosphate (PO4), it makes up an important part of the structural framework that holds DNA and RNA together. Phosphates are also a critical component of ATP—the cellular energy carrier—as they serve as an energy ?release' for organisms to use in building proteins or contacting muscles. Like calcium, phosphorus is important to vertebrates; in the human body, 80% of phosphorous is found in teeth and bones.

The phosphorus cycle differs from the other major biogeochemical cycles in that it does not include a gas phase; although small amounts of phosphoric acid (H3PO4) may make their way into the atmosphere, contributing—in some cases—to acid rain. The water, carbon, nitrogen and sulfur cycles all include at least one phase in which the element is in its gaseous state. Very little phosphorus circulates in the atmosphere because at Earth's normal temperatures and pressures, phosphorus and its various compounds are not gases. The largest reservoir of phosphorus is in sedimentary rock.

It is in these rocks where the phosphorus cycle begins. When it rains, phosphates are removed from the rocks (via weathering) and are distributed throughout both soils and water. Plants take up the phosphate ions from the soil. The phosphates then moves from plants to animals when herbivores eat plants and carnivores eat plants or herbivores. The phosphates absorbed by animal tissue through consumption eventually returns to the soil through the excretion of urine and feces, as well as from the final decomposition of plants and animals after death.

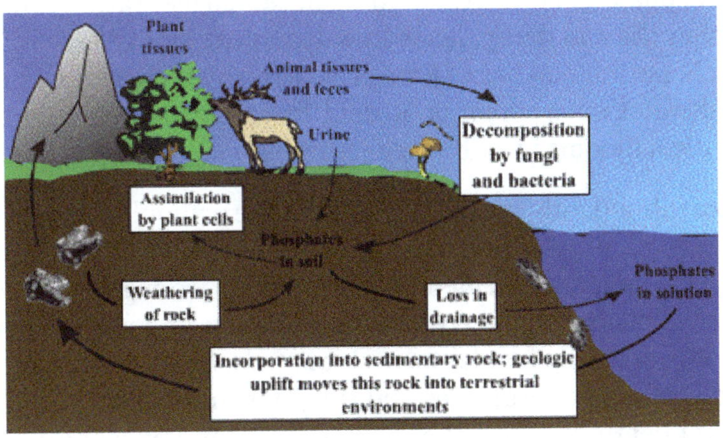

The same process occurs within the aquatic ecosystem. Phosphorus is not highly soluble, binding tightly to molecules in soil, therefore it mostly reaches waters by traveling with runoff soil particles. Phosphates also enter waterways through fertilizer runoff, sewage seepage, natural mineral deposits, and wastes from other industrial processes. These phosphates tend to settle on ocean floors and lake bottoms. As sediments are stirred up, phosphates may reenter the phosphorus cycle, but they are more commonly made available to aquatic organisms by being exposed through erosion. Water plants take up the waterborne phosphate which then travels up through successive stages of the aquatic food chain.

While obviously beneficial for many biological processes, in surface waters an excessive concentration of phosphorus is considered a pollutant. Phosphate stimulates the growth of plankton and plants, favoring weedy species over others. Excess growth of these plants tend to consume large amounts of dissolved oxygen, potentially suffocating fish and other marine animals, while also blocking available sunlight to bottom dwelling species. This is known as eutrophication.

Phosphorus in the Environment

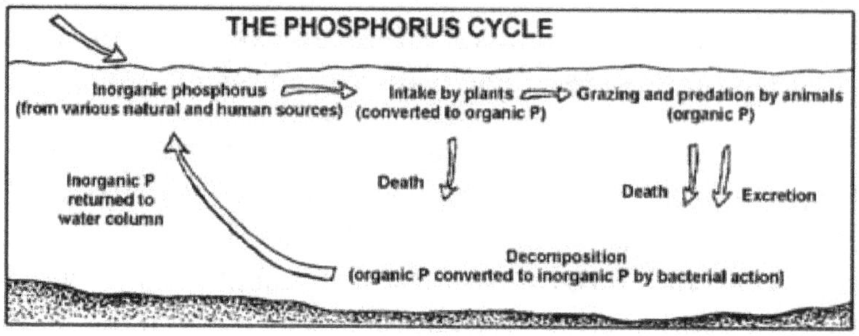

The aquatic phosphorus cycle.

Ecological Function

Phosphorus is an essential nutrient for plants and animals. Phosphorus is a limiting nutrient for aquatic organisms. Phosphorus forms parts of important life-sustaining molecules that are very common in the biosphere. Phosphorus does not enter the atmosphere, remaining mostly on land and in rock and soil minerals. Eighty percent of the mined phosphorus is used to make fertilizers. Phosphates from fertilizers, sewage and detergents can cause pollution in lakes and streams. Overenrichment of phosphate in both fresh and inshore marine waters can lead to massive algae blooms which, when they die and decay, leads to eutrophication of fresh waters only. An example of this is the Canadian Experimental Lakes Area. These freshwater algal blooms should not be confused with those in saltwater environments. Recent research suggests that the predominant pollutant responsible for algal blooms in salt water estuaries and coastal marine habitats is Nitrogen.

Phosphorus occurs most abundantly in nature as part of the orthophosphate ion $(PO_4)^{3-}$, consisting of a P atom and 4 oxygen atoms. On land most phosphorus is found in rocks and minerals. Phosphorus rich deposits have generally formed in the ocean or from guano, and over time, geologic processes bring ocean sediments to land. Weathering of rocks and minerals release phosphorus in a soluble form where it is taken up by plants, and it is transformed into organic compounds. The plants may then be consumed by herbivores and the phosphorus is either incorporated into their tissues or excreted. After death, the animal or plant decays, and phosphorus is returned to the soil where a large part of the phosphorus is transformed into insoluble compounds. Runoff may carry a small part of the phosphorus back to the ocean. Generally with time (thousands of years) soils become deficient in phosphorus leading to ecosystem retrogression.

Biological Function

The primary biological importance of phosphates is as a component of nucleotides, which serve as energy storage within cells (ATP) or when linked together, form the nucleic acids DNA and RNA. The double helix of our DNA is only possible because of the phosphate ester bridge that binds the helix. Besides making biomolecules, phosphorus is also found in bone and the enamel of mammalian teeth, whose strength is derived from calcium phosphate in the form of Hydroxylapatite. It is also found in the exoskeleton of insects, and phospholipids (found in all biological membranes). It also functions as a buffering agent in maintaining acid base homeostasis in the human body.

Process of the Cycle

Phosphates move quickly through plants and animals; however, the processes that move them through the soil or ocean are very slow, making the phosphorus cycle overall one of the slowest biogeochemical cycles.

Initially, phosphate weathers from rocks and minerals, the most common mineral being apatite. Overall small losses occur in terrestrial environments by leaching and erosion, through the action of rain. In soil, phosphate is adsorbed on iron oxides, aluminium hydroxides, clay surfaces, and organic matter particles, and becomes incorporated (immobilized or fixed). Plants and fungi can also be active in making P soluble.

Unlike other cycles, P is not normally found in the air as a gas; it only occurs under highly reducing conditions as the gas phosphine PH_3.

Phosphatic Minerals

The availability of phosphorus in an ecosystem is restricted by the rate of release of this element during weathering. The release of phosphorus from apatite dissolution is a key control on ecosystem productivity. The primary mineral with significant phosphorus content, apatite $[Ca_5(PO_4)_3OH]$ undergoes carbonation.

Little of this released phosphorus is taken up by biota (organic form), whereas a larger proportion reacts with other soil minerals. This leads to precipitation into unavailable forms in the later stage of weathering and soil development. Available phosphorus is found in a biogeochemical cycle in the upper soil profile, while phosphorus found at lower depths is primarily involved in geochemical reactions with secondary minerals. Plant growth depends on the rapid root uptake of phosphorus released from dead organic matter in the biochemical cycle. Phosphorus is limited in supply for plant growth. Phosphates move quickly through plants and animals; however, the processes that move them through the soil or ocean are very slow, making the phosphorus cycle overall one of the slowest biogeochemical cycles.

Low-molecular-weight (LMW) organic acids are found in soils. They originate from the activities of various microorganisms in soils or may be exuded from the roots of living plants. Several of those organic acids are capable of forming stable organo-metal complexes with various metal ions found in soil solutions. As a result, these processes may lead to the release of inorganic phosphorus associated with aluminium, iron, and calcium in soil minerals. The production and release of oxalic acid by mycorrhizal fungi explain their importance in maintaining and supplying phosphorus to plants.

The availability of organic phosphorus to support microbial, plant and animal growth depends on the rate of their degradation to generate free phosphate. There are various enzymes such as phosphatases, nucleases and phytase involved for the degradation. Some of the abiotic pathways in the environment studied are hydrolytic reactions and photolytic reactions. Enzymatic hydrolysis of organic phosphorus is an essential step in the biogeochemical phosphorus cycle, including the phosphorus nutrition of plants and microorganisms and the transfer of organic phosphorus from soil to bodies of water. Many organisms rely on the soil derived phosphorus for their phosphorus nutrition.

Human Influences

Nutrients are important to the growth and survival of living organisms, and hence, are essential for development and maintenance of healthy ecosystems. Humans have greatly influenced the phosphorus cycle by mining phosphorus, converting it to fertilizer, and by shipping fertilizer and products around the globe. Transporting phosphorus in food from farms to cities has made a major change in the global Phosphorus cycle. However, excessive amounts of nutrients, particularly phosphorus and nitrogen, are detrimental to aquatic ecosystems. Waters are enriched in phosphorus from farms' run-off, and from effluent that is inadequately treated before it is discharged to waters. Natural eutrophication is a process by which lakes gradually age and become more productive and may take thousands of years to progress. Cultural or anthropogenic eutrophication, however, is water pollution caused by excessive plant nutrients; this results in excessive growth in the algal population; when this algae dies its putrefaction depletes the water of oxygen. Such eutrophication may also give rise to toxic algal bloom. Both these effects cause animal and plant death rates to increase as the plants take in poisonous water while the animals drink the poisoned water. Surface and subsurface runoff and erosion from high-phosphorus soils may be major contributing factors to this fresh water eutrophication. The processes controlling soil Phosphorus release to surface runoff and to subsurface flow are a complex interaction between the type of phosphorus input, soil type and management, and transport processes depending on hydrological conditions.

Repeated application of liquid hog manure in excess to crop needs can have detrimental effects on soil phosphorus status. Also, application of biosolids may increase available phosphorus in soil. In poorly drained soils or in areas where snowmelt can cause periodic waterlogging, dereducing conditions can be attained in 7–10 days. This causes a sharp increase in phosphorus concentration in solution and phosphorus can be leached. In addition, reduction of the soil causes a shift in phosphorus from resilient to more labile forms. This could eventually increase the potential for phosphorus loss. This is of particular concern for the environmentally sound management of such areas, where disposal of agricultural wastes has already become a problem. It is suggested that the water regime of soils that are to be used for organic wastes disposal is taken into account in the preparation of waste management regulations.

Human interference in the phosphorus cycle occurs by overuse or careless use of phosphorus fertilizers. This results in increased amounts of phosphorus as pollutants in bodies of water resulting in eutrophication. Eutrophication devastates water ecosystems by inducing anoxic conditions.

Sulfur Cycle

Sulfur (S), the tenth most abundant element in the universe, is a brittle, yellow, tasteless, and odorless non-metallic element. It comprises many vitamins, proteins, and hormones that play critical roles in both climate and in the health of various ecosystems. The majority of the Earth's sulfur is stored underground in rocks and minerals, including as sulfate salts buried deep within ocean sediments.

The sulfur cycle contains both atmospheric and terrestrial processes. Within the terrestrial portion, the cycle begins with the weathering of rocks, releasing the stored sulfur. The sulfur then comes into contact with air where it is converted into sulfate (SO4). The sulfate is taken up by plants and microorganisms and is converted into organic forms; animals then consume these organic forms through foods they eat, thereby moving the sulfur through the food chain. As organisms die and decompose, some of the sulfur is again released as a sulfate and some enters the tissues of microorganisms. There are also a variety of natural sources that emit sulfur directly into the atmosphere, including volcanic eruptions, the breakdown of organic matter in swamps and tidal flats, and the evaporation of water.

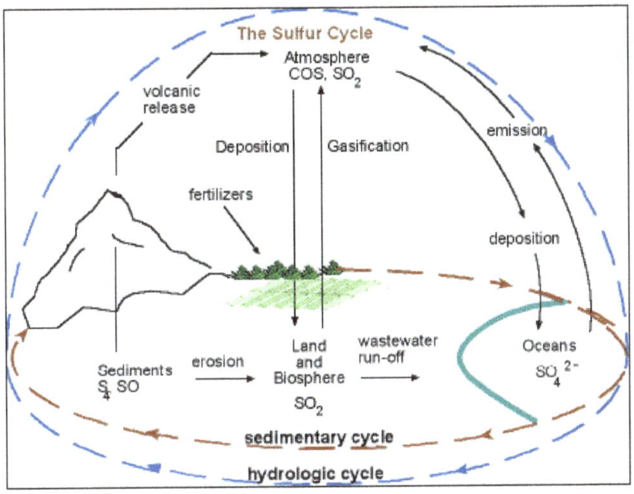

Sulfur eventually settles back into the Earth or comes down within rainfall. A continuous loss of sulfur from terrestrial ecosystem runoff occurs through drainage into lakes and streams, and eventually oceans. Sulfur also enters the ocean through fallout from the Earth's atmosphere. Within the ocean, some sulfur cycles through marine communities, moving through the food chain. A portion of this sulfur is emitted back into the atmosphere fromd sea spray. The remaining sulfur is lost to the ocean depths, combining with iron to form ferrous sulfide which is responsible for the black color of most marine sediments.

Since the Industrial Revolution, human activities have contributed to the amount of sulfur that enters the atmosphere, primarily through the burning of fossil fuels and the processing of metals. One-third of all sulfur that reaches the atmosphere—including 90% of sulfur dioxide—stems from human activities. Emissions from these activities, along with nitrogen emissions, react with other chemicals in the atmosphere to produce tiny particles of sulfate salts which fall as acid rain, causing a variety of damage to both the natural environment as well as to man-made environments, such as the chemical weathering of buildings. However, as particles and tiny airborne droplets, sulfur also acts as a regulator of global climate. Sulfur dioxide and sulfate aerosols absorb ultraviolet radiation, creating cloud cover that cools cities and may offset global warming caused by the greenhouse effect. The actual amount of this offset is a question that researchers are attempting to answer.

Most of the earth's sulphur is tied up in rocks and salts or buried deep in the ocean in oceanic sediments. Sulphur can also be found in the atmosphere. It enters the atmosphere through both natural and human sources. Natural recourses can be for instance volcanic eruptions, bacterial processes, evaporation from water, or decaying organisms. When sulphur enters the atmosphere through human activity, this is mainly a consequence of industrial processes where sulphur dioxide (SO_2) and hydrogen sulphide (H_2S) gases are emitted on a wide scale. When sulphur dioxide enters the atmosphere it will react with oxygen to produce sulphur trioxide gas (SO_3), or with other chemicals in the atmosphere, to produce sulphur salts. Sulphur dioxide may also react with water to produce sulphuric acid (H_2SO_4). Sulphuric acid may also be produced from demethylsulphide, which is emitted to the atmosphere by plankton species. All these particles will settle back onto earth, or react with rain and fall back onto earth as acid deposition. The particles will than be absorbed by plants again and are released back into the atmosphere, so that the sulphur cycle will start over again.

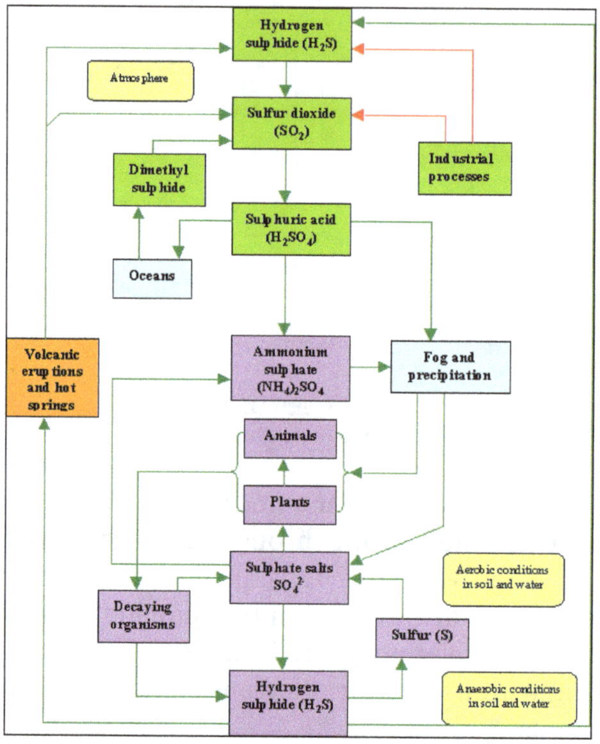

A schematic representation of the sulphur cycle.

References

- Marchant, H. K., Lavik, G., Holtappels, M., and Kuypers, M. M. M. (2014). "The Fate of Nitrate in Intertidal Permeable Sediments". PLoS ONE. 9 (8). doi:10.1371/journal.pone.0104517
- McGuire, 1A. D.; Lukina, N. V. (2007). "Biogeochemical cycles". In Groisman, P.; Bartalev, S. A.; NEESPI Science Plan Development Team. Northern Eurasia earth science partnership initiative (NEESPI), Science plan overview (PDF). Global Planetary Change. 56. pp. 215–234. Retrieved 09, June 2020
- "Mercury Cycling in the Environment". Wisconsin Water Science Center. United States Geological Survey. 10 January 2013. Retrieved 11, March 2020
- Merrill, Eugene H.; Rooker, Mark F.; Grisanti, Michael A. (2011). The World and the Word. Nashville, TN: B&H Academic. p. 430. ISBN 9780805440317
- Paul J. Durack; Susan E. Wijffels & Richard J. Matear (27 April 2012). "Ocean Salinities Reveal Strong Global Water Cycle Intensification During 1950 to 2000". Science. 336 (6080): 455–458. Bibcode:2012Sci...336..455D. doi:10.1126/science.1212222. PMID 22539717

- Baedke, Steve J.; Fichter, Lynn S. "Biogeochemical Cycles: Carbon Cycle". Supplimental Lecture Notes for Geol 398. James Madison University. Retrieved 22, July 2020
- Lal, Rattan (2008). "Sequestration of atmospheric CO_2 in global carbon pools". Energy and Environmental Science. 1: 86–100. doi:10.1039/b809492f
- Britto, Dev T.; Kronzucker, Herbert J. "NH4+ toxicity in higher plants: a critical review". Journal of Plant Physiology. 159 (6): 567–584. doi:10.1078/0176-1617-0774
- Rudolf Dvořák (2007). Extrasolar Planets. Wiley-VCH. pp. 139–140. ISBN 978-3-527-40671-5. Retrieved 21, January 2020

Permissions

All chapters in this book are published with permission under the Creative Commons Attribution Share Alike License or equivalent. Every chapter published in this book has been scrutinized by our experts. Their significance has been extensively debated. The topics covered herein carry significant information for a comprehensive understanding. They may even be implemented as practical applications or may be referred to as a beginning point for further studies.

We would like to thank the editorial team for lending their expertise to make the book truly unique. They have played a crucial role in the development of this book. Without their invaluable contributions this book wouldn't have been possible. They have made vital efforts to compile up to date information on the varied aspects of this subject to make this book a valuable addition to the collection of many professionals and students.

This book was conceptualized with the vision of imparting up-to-date and integrated information in this field. To ensure the same, a matchless editorial board was set up. Every individual on the board went through rigorous rounds of assessment to prove their worth. After which they invested a large part of their time researching and compiling the most relevant data for our readers.

The editorial board has been involved in producing this book since its inception. They have spent rigorous hours researching and exploring the diverse topics which have resulted in the successful publishing of this book. They have passed on their knowledge of decades through this book. To expedite this challenging task, the publisher supported the team at every step. A small team of assistant editors was also appointed to further simplify the editing procedure and attain best results for the readers.

Apart from the editorial board, the designing team has also invested a significant amount of their time in understanding the subject and creating the most relevant covers. They scrutinized every image to scout for the most suitable representation of the subject and create an appropriate cover for the book.

The publishing team has been an ardent support to the editorial, designing and production team. Their endless efforts to recruit the best for this project, has resulted in the accomplishment of this book. They are a veteran in the field of academics and their pool of knowledge is as vast as their experience in printing. Their expertise and guidance has proved useful at every step. Their uncompromising quality standards have made this book an exceptional effort. Their encouragement from time to time has been an inspiration for everyone.

The publisher and the editorial board hope that this book will prove to be a valuable piece of knowledge for students, practitioners and scholars across the globe.